PLANT ECOGENOMICS

PLANT ECOGENOMICS

Edited by
Peerzada Arshid Shabir, PhD
Peerzada Yasir Yousuf, PhD
Khalid Rehman Hakeem, PhD

AAP | APPLE ACADEMIC PRESS

First edition published 2023

Apple Academic Press Inc.
1265 Goldenrod Circle, NE,
Palm Bay, FL 32905 USA

760 Laurentian Drive, Unit 19,
Burlington, ON L7N 0A4, CANADA

CRC Press
6000 Broken Sound Parkway NW,
Suite 300, Boca Raton, FL 33487-2742 USA

4 Park Square, Milton Park,
Abingdon, Oxon, OX14 4RN UK

Library and Archives Canada Cataloguing in Publication

Title: Plant ecogenomics / edited by Peerzada Arshid Shabir, PhD, Peerzada Yasir Yousuf, PhD, Khalid Rehman Hakeem, PhD.

Names: Shabir, Peerzada Arshid, editor. | Yousuf, Peerzada Yasir, editor. | Hakeem, Khalid Rehman, editor.

Description: First edition. | Includes bibliographical references and index.

Identifiers: Canadiana (print) 20220255733 | Canadiana (ebook) 20220255776 | ISBN 9781774910184 (hardcover) | ISBN 9781774910191 (softcover) | ISBN 9781003282006 (ebook)

Subjects: LCSH: Plant genetics. | LCSH: Plant ecology. | LCSH: Metagenomics.

Classification: LCC QK981 .P53 2023 | DDC 581.3/5—dc23

Library of Congress Cataloging-in-Publication Data

Names: Shabir, Peerzada Arshid, editor. | Yousuf, Peerzada Yasir, editor. | Hakeem, Khalid Rehman, editor.

Title: Plant ecogenomics / edited by Peerzada Arshid Shabir, Peerzada Yasir Yousuf, Khalid Rehman Hakeem

Description: First edition | Palm Bay, FL, USA : Apple Academic Press, 2023. | Includes bibliographical references and index. | Summary: "Plant Ecogenomics offers a valuable introduction to plant ecology from a genomics point of view. The editors present a thorough foundation and summary of modern approaches, methodologies, research goals, and evidence of plant ecology in the modern genomic era. They also present important updated information on the most recent knowledge of the different aspects of plant populations. The volume begins with a general overview of the various approaches to unravelling the genetic basis underlying fundamental responses of plants to their natural environments. It also describes in detail the concepts, aims, and approaches of plant ecological genomics along with the genomic tools embraced by ecologists for mining various ecological problems. Recent advances and breakthroughs made in molecular markers along with their applications in plant ecogenomic studies are shared, and specific applications, techniques, and tools are described as well, such advanced molecular techniques, next-generation sequencing, eDNA metabarcoding, among others. The authors look at community and ecosystem responses in the form of phylogenetic systematics and the theory and methodology for inferring phylogenetic trees or cladograms, including parsimony, Bayesian, and maximum likelihood methods. The employability of QTL analysis and reverse genetics in plant ecogenomics is discussed. Other topics include the role of eco-genomics in conservation and management, models to study the population structure along with the genetic basis of speciation and adaptation in plants, and more. The volume looks at how genetic variability is assessed with the help of various available tools and how effective they are in the conservation of important plant species. Inbreeding and genetic load and the role of ecogenomics in landscape genetic restoration are covered as well. This volume offers a rich store of information on the interdisciplinary field of ecological genomics that will be valuable for molecular biologists, ecologists, as well as for researchers, scientists, and faculty and students in related areas"-- Provided by publisher.

Identifiers: LCCN 2022025310 (print) | LCCN 2022025311 (ebook) | ISBN 9781774910184 (hardback) | ISBN 9781774910191 (paperback) | ISBN 9781003282006 (ebook)

Subjects: LCSH: Plant genetics. | Plant ecology. | Metagenomics.

Classification: LCC QK981 .P48 2023 (print) | LCC QK981 (ebook) | DDC 581.3/5--dc23/eng/20220707

LC record available at https://lccn.loc.gov/2022025310

LC ebook record available at https://lccn.loc.gov/2022025311

ISBN: 978-1-77491-018-4 (hbk)
ISBN: 978-1-77491-019-1 (pbk)
ISBN: 978-1-00328-200-6 (ebk)

About the Editors

Peerzada Arshid Shabir, PhD, MSc, is a plant science professional with a specialization in plant reproductive ecology and is currently working as Assistant Professor of Botany in the Department of Higher Education, Jammu and Kashmir, India. Dr. Shabir received his PhD in the field of plant reproductive ecology and genetic diversity from the University of Kashmir, Srinagar. He has published several research articles and book chapters in reputed international journals and has also presented his work at various national and international conferences. He has also been a reviewer for many reputed international journals. His current research interests include the genomic bases of phenotypic plasticity and the study of the genomic bases of adaptation and speciation. He has ten years of research experience and is currently actively engaged in the teaching of undergraduate subjects like plant reproductive ecology, plant physiology, cell and molecular biology, and genetic engineering of plants. Dr. Shabir has also been conferred with junior and senior research fellowships by the Counsel of Scientific and Industrial Research, New Delhi.

Peerzada Yasir Yousuf, PhD, is Assistant Professor of Botany in the Department of Higher Education, Government of Jammu and Kashmir, India. He has published several research articles and book chapters in globally reputed journals, including *Frontiers in Plant Science, Genes, Plos One, Protoplasma, Plant Growth Regulation, Environmental Science and Pollution Research*, etc. Dr. Yousuf has presented his work at several national and international conferences. He has also delivered extension lectures in different fields of plant sciences at various institutes. He is a lifetime member of several national and international organizations and is the recipient of several fellowships. Dr. Yousuf also serves as a reviewer and editorial board member of many research journals. He is actively engaged in the research fields of plant physiology and molecular ecology. He formerly worked at the Molecular Ecology Laboratory as a Senior Research Fellow in the Department of Botany Jamia Hamdard, New Delhi. Dr. Yousuf also taught environmental science as guest faculty in Jamia Hamdard in different faculties, including Faculty of Unani Medicine and the Faculty of Information Technology. He has teaching experience of 11 years and research experience of eight years. Dr. Yousuf earned his BSc from the University of Kashmir and MSc and PhD degrees in Botany from Jamia Hamdard, New Delhi, respectively. During his PhD program, he worked on the biochemical and molecular aspects of stress tolerance in plants. He has qualified for many national competitive examinations like UGC-CSIR NET (AIR 39), ARS-NET, GATE (AIR 45), and JK-SET. In addition, he has completed many online certification courses in the various fields of biology from different nationally and internationally reputed institutes. He was selected for the SERB National Postdoctoral Fellowship (N-PDF) by the Department of Science and Technology, Government of India in 2016.

Khalid Rehman Hakeem, PhD, is Professor at King Abdulaziz University, Jeddah, Saudi Arabia. After completing his doctorate (botany, specialization in plant eco-physiology and molecular biology) from Jamia Hamdard, New Delhi, India, he worked as a lecturer at the University of Kashmir, Srinagar, India, at Universiti Putra Malaysia, Selangor, Malaysia, where he was a Postdoctorate Fellow and Fellow Researcher (Associate Professor) for several years. Dr. Hakeem has more than 10 years of teaching and research experience in plant eco-physiology, biotechnology and molecular biology, medicinal plant research, plant-microbe-soil interactions, as well as in environmental studies. He is the recipient of several fellowships at both national and international levels. He has served as a visiting scientist at Jinan University, Guangzhou, China. Currently, he is involved with several international research projects with various government organizations. To date, Dr. Hakeem has authored and edited more than 60 books with international publishers. He also has to his credit more than 115 research publications in peer-reviewed international journals and 55 book chapters in edited volumes. At present, Dr. Hakeem serves as an editorial board member and reviewer for several high-impact international scientific journals. He is included in the advisory board of Cambridge Scholars Publishing, UK. He is also a fellow of the Plantae Group of the American Society of Plant Biologists, member of the World Academy of Sciences, member of the International Society for Development and Sustainability, Japan, and member of the Asian Federation of Biotechnology, Korea.

Contents

Contributors

Basharat Ahmad Bhat
Department of Bioresources, School of Biological Sciences, University of Kashmir, Srinagar, Jammu and Kashmir, India

Fayaz Ahmad Dar
Department of Bioresources, Amar Singh College Campus, Cluster University, Srinagar 190008, Jammu and Kashmir, India; E-mail: dfayaz5u@gmail.com

Tanvir Ul Hassan Dar
Department of Biotechnology, School of Biosciences and Biotechnology, BGSB University, Rajouri, Jammu and Kashmir, India; E-mail: tanvirulhasan@gmail.com

Meenu Garg
Department of Microbiology, Shaheed Rajguru College of Applied Sciences for Women, University of Delhi, 110096 Delhi, India

Aflaq Hamid
Division of Plant Pathology, Sher-e-Kashmir University of Agricultural Science and Technology, Shalimar Srinagar, Kashmir, India; E-mail: falak19@gmail.com

Khalid Rehman Hakeem
Department of Biological Sciences, Faculty of Science, KAU, Jeddah, Saudi Arabia; E-mail: kur.hakeem@gmail.com

Shahid ul Islam
Department of Biotechnology, School of Biosciences and Biotechnology, BGSB University, Rajouri, Jammu and Kashmir, India

Sheikh Tajamul Islam
Department of Bioresources, School of Biological Sciences, University of Kashmir, Srinagar, Jammu and Kashmir, India

Sabitri Kumari
Laboratory of Morphogenesis, Centre of Advanced Study in Botany, Institute of Science, Banaras Hindu University, Varanasi 221005, Uttar Pradesh, India

Mahak Majeed
Department of Bioresources, University of Kashmir, Hazratbal, Srinagar, Jammu and Kashmir, India

Zahid Ahmad Mangral
Department of Biotechnology, School of Biosciences and Biotechnology, BGSB University, Rajouri, Jammu and Kashmir, India

Bipin Maurya
Department of Botany, Institute of Science, Banaras Hindu University, Varanasi 221005, Uttar Pradesh, India; E-mail: bipinmaurya841@gmail.com
Laboratory of Morphogenesis, Centre of Advanced Study in Botany, Institute of Science, Banaras Hindu University, Varanasi 221005, Uttar Pradesh, India

Haider Mehraj
Department of Electronics and Communication Engineering, School of Engineering and Technology, BGSB University, Rajouri, Jammu and Kashmir, India

Tarun Minocha
Department of Zoology, Institute of Science, Banaras Hindu University, Varanasi 2210005, India; E-mail: tarunbhu89@gmail.com

Naveed ul Mushtaq
Department of Bioresources, School of Biological Sciences, University of Kashmir, Srinagar 190006, India

Adil Rashid Nath
Division of Plant Biotechnology, Sher-e-Kashmir University of Agricultural Science and Technology, Shalimar Srinagar, Kashmir, India

Pragyanshree Nayak
Department of Microbiology, Shaheed Rajguru College of Applied Sciences for Women, University of Delhi, Delhi 110096, India

Divya Raj Prasad
Department of Genetics and Plant Breeding, Institute of Agricultural Sciences, Banaras Hindu University, Varanasi 2210005, India; E-mail: divyarajbhu@gmail.com

Nidhi Rai
Laboratory of Morphogenesis, Centre of Advanced Study in Botany, Institute of Science, Banaras Hindu University, Varanasi 221005, Uttar Pradesh, India

Shashi Pandey-Rai
Laboratory of Morphogenesis, Centre of Advanced Study in Botany, Institute of Science, Banaras Hindu University, Varanasi 221005, Uttar Pradesh, India; E-mail: Shashi.bhubotany@gmail.com

Ashwini Kumar Ray
Department of Microbiology, Shaheed Rajguru College of Applied Sciences for Women, University of Delhi, Delhi 110096, India; E-mail: ashwinires21@gmail.com

Shahjahan Rashid
Division of Plant Pathology, Sher-e-Kashmir University of Agricultural Science and Technology, Shalimar Srinagar, Kashmir, India

Aadil Rasool
Department of Bioresources, School of Biological Sciences, University of Kashmir, Srinagar 190006, India

Reiaz Ul Rehman
Department of Bioresources, School of Biological Sciences, University of Kashmir, Srinagar 190006, India; E-mail: rreiazbiores@gmail.com

Seerat Saleem
Department of Bioresources, School of Biological Sciences, University of Kashmir, Srinagar 190006, India

Lakee Sharma
Laboratory of Morphogenesis, Centre of Advanced Study in Botany, Institute of Science, Banaras Hindu University, Varanasi 221005, Uttar Pradesh, India

Wasifa Hafiz Shah
Department of Bioresources, School of Biological Sciences, University of Kashmir, Srinagar 190006, India

Alok Kumar Singh
Department of Biochemistry, Shaheed Rajguru College of Applied Sciences for Women, University of Delhi, Delhi 110096, India

Aparna Singh
Department of Plant Sciences, Avvaiyar Government College for Women, Karaikal 609602, Puducherry, India; E-mail: aparnasingh@dhtepdy.edu.in

Harshit Kumar Soni
Department of Zoology, Government Science College, Pandhurna 480334, Madhya Pradesh, India; E-mail: hksoni2003@gmail.com

Inayatullah Tahir
Department of Bioresources, School of Biological Sciences, University of Kashmir, Srinagar, Jammu and Kashmir, India

Lubna Tariq
Department of Biotechnology, School of Biosciences and Biotechnology, BGSB University, Rajouri, Jammu and Kashmir, India

Bilal Ahmad Wani
Department of Botany, University of Kashmir, Jammu and Kashmir, India

Farhana Wani
Division of Plant Pathology, Sher-e-Kashmir University of Agricultural Science and Technology, Shalimar Srinagar, Kashmir, India

Sumiah Wani
Division of Plant Pathology, Sher-e-Kashmir University of Agricultural Science and Technology, Shalimar Srinagar, Kashmir, India

Peerzada Yasir Yousuf
Department of Botany, Government Degree College Pulwama Jammu and Kashmir India; E-mail: Syedyasar55@gmail.com

Abbreviations

AFLP	amplified fragment length polymorphism
AM	arbuscular mycorrhizae
BPA	bisphenol A
CAPS	cleaved amplified polymorphic sequences
CBOL	Consortium for the Barcode of Life
CCD	charge-coupled device
CLR	continuous long reads
CNV	copy number variation
cpSSRs	chloroplast microsatellites
DArT	diversity arrays technology
DMRs	differentially methylated regions
eDNA	environmental DNA
EST	expressed sequence tag
FER	*Fusarium* ear rot
FRET	fluorescence resonance energy transfer
FST	Wright's F-statistics
GEAs	genetic-environment associations
GBS	genotyping by sequencing
GR	glucocorticoid receptor
gSSRs	genomic simple sequenced repeats
GWAS	genome-wide association study
GWSS	genome-wide selection scan
HTS	high-throughput sequencing
ISSR	inter-simple sequence repeats
IRAP	inter-retrotransposon amplified polymorphism
ITS	internal transcribed spacer
MAS	marker-assisted selection
MBTs	molecular biology tools
MPSS	massively parallel signature sequencing
MS-AFLP	methyl-sensitive amplified fragment length polymorphism
mtDNA	mitochondrial DNA
NGS	next-generation sequencing

NJ	neighbor-joining
ORFs	open reading frames
PacBio	Pacific Biosciences
PCR	polymerase chain reaction
PGM	Personal Genome Machine
PGRs	plant genetic resources
PTGS	posttranscriptional gene silencing
QTL	quantitative trait locus
RAMP	randomly amplified microsatellite polymorphism
RAPD	random amplified polymorphic DNA
RBIP	retrotransposon-based insertion polymorphism
REMAP	retrotransposon microsatellite amplified polymorphism
RFLP	restriction fragment length polymorphism
RIL	recombinant inbred lines
RNAi	RNA interference
SBL	sequencing by ligation
SBS	sequencing by synthesis
SCAR	sequence characterized amplified regions
SCoT	start codon targeted
siRNA	small interfering RNA
SMRT	single-molecule real-time
SNP	single nucleotide polymorphisms
SOLiD	supported oligonucleotide ligation and detection
SRAP	sequence-related amplified polymorphism
SRM	selected reaction monitoring
sRNAs	small RNAs
S-SAP	sequence-specific amplification polymorphism
SSCP	single-strand conformation polymorphism
SSR	simple sequence repeat
TEM	tunneling electron microscopy
TERN	Terrestrial Ecosystem Research Network
TEs	transposable elements
TRAP	target region amplification polymorphism
tSMS	true single-molecule sequencing
UPGMA	unweighted pair group method with arithmetic mean
UV	ultraviolet
ZMW	zero-mode waveguide

Foreword

Ecogenomics aims to recognize the mechanisms of genes and genetic mechanism that underlie the fundamental responses of organisms to their natural environs. It crosses disciplines to determine the extent to which those genes and pathways exhibit functional variation in nature and to characterize the ecological and evolutionary consequences of that variation. Achieving this aim will inevitably require a multidisciplinary approach, and the rapid technological advances in genomics methods and bioinformatics tools have made interdisciplinary collaborations decisive for the successful implementation of ecogenomics studies. However, integration across disciplines is far from a simple task, as we not only require expertise in different experimental areas, but also a conceptual integration and understanding of mechanisms and interactions at different levels of biological organization are also needed.

In this book, methodologies that are established and emerging in ecological genomics have been discussed, and how this framework can address long-standing and emerging questions in plant ecology is emphasized. The present book aims to highlight how indispensable the ecogenomics approach is and how the advances in the field of genomics have revolutionized the field of ecology.

The information is presented across 12 different chapters. The first half of the book provides a brief history of plant ecogenomics followed by an overview of different molecular techniques and approaches employed in plant ecogenomics and the types of data they generate. This leads to an overview of population genetics in ecology. The second half of the book introduces the essential underpinnings of molecular ecology and describes in detail the recent molecular techniques that are nowadays frequently used in ecological research. In the last section, the book explains specific applications of ecogenomics covering conservation genetics and genetic restoration.

Prof. Munir Ozturk
(Prof. Emeritus) Department of Botany
Ege University, Izmir, Turkey

Preface

Plant Ecogenomics is an introduction to plant ecology from a genomics point of view. Our objective is to present a foundation of the modern approaches, methodologies, research goals, and shreds of evidence of plant ecology in the modern genomic era and to summarize the most recent knowledge of the different aspects of plant populations. We have tried to present the material in a condensed but clear manner, so that readers can better understand the important parts of the voluminous information in the field. The book is meant to serve related researchers and students at the college graduate and upper-undergraduate levels in plant ecology and molecular biology. Numerous tables, line drawings, and color photographs are used throughout. A key feature is that the recent references for each chapter have been listed to incorporate the updates. The literature cited in the references is not exhaustive, so students are encouraged to do literature searches on their own.

The book is classified into 12 chapters so that the reader may choose what to read, depending on personal preference. **Chapter 1**, *Ecogenomics: An Appraisal*, gives a general overview of the various approaches to unravel the genetic basis underlying the fundamental responses of plants to their natural environments. In addition, it also describes in detail the concepts, aims, and approaches of plant ecological genomics along with the genomic tools embraced by ecologists for mining various ecological problems. **Chapter 2** serves as an introduction to the advances made in the molecular markers along with their applications in plant ecogenomic studies. **Chapter 3** covers the details of the advanced molecular technique and next-generation sequencing, briefs how new trends in the field of sequencing can shape the future of omics, and studies of non-model organisms in an ecological context. **Chapter 4** discusses the various potential attributes and recent advancements achieved in eDNA metabarcoding approaches applied to various plant taxonomic groups. **Chapter 5** determines the community and ecosystem responses in the form of phylogenetic systematics and the theory and methodology for inferring phylogenetic trees or cladograms, including parsimony, Bayesian, and maximum likelihood methods.

The employability of QTL Analysis and reverse genetics in plant ecogenomics have been discussed in **Chapter 6**. In **Chapter 7**, different dispersal modes, bottleneck effect on gene flow, and the currently available methods and models to study the population structure have been thoroughly examined. The genetic basis of speciation and adaptation in plants has been thoroughly discussed in **Chapter 8**. The focus of **Chapter 9** is on the role of ecogenomics in conservation and management. In **Chapter 10**, the focus has been on how the genetic variability is assessed with the help of various available tools and how effective they are in the conservation of important plant species. **Chapter 11** discusses the inbreeding and genetic load and role of ecogenomics in landscape genetic restoration. Lastly, **Chapter 12** provides an overview of plant ecogenomics and upcoming challenges and opportunities.

— Editors

CHAPTER 1

Ecogenomics: An Appraisal

BILAL AHMAD WANI[1], PEERZADA YASIR YOUSUF[2], and
PEERZADA ARSHID SHABIR[3,*]

[1]*Department of Botany, University of Kashmir 190006, Jammu and Kashmir, India*

[2]*Department of Botany, Govt. Degree College Pulwama 192301, Jammu and Kashmir, India*

[3]*Department of Botany, Govt. Degree College Sopore 193201, Jammu and Kashmir, India*

[*]*Corresponding author. E-mail: peerzadarshid@gmail.com*

ABSTRACT

Ecological genomics integrates genomic and ecological approaches to unravel the genetic basis underlying biochemical, physiological, morphological, or behavioral responses of organisms to the environment. The discipline aims to recognize the genetic mechanism underlying to fundamental responses of organisms to their natural environs. Functional genomic approaches are being employed to classify and characterize genes of ecological and evolutionary significance. At present, molecular biology methods and techniques are gradually being embraced by plant ecologists to understand ecological genomics and to examine the mechanistic and evolutionary base point of species interaction. The modern field of Ecogenomics seeks to realize the mechanistic basis for adaptation and phenotypic variations and focus to pinpoint the genes influenced by

Plant Ecogenomics. Peerzada Arshid Shabir, PhD, Peerzada Yasir Yousuf, PhD &
Khalid Rehman Hakeem, PhD (Eds.)

evolution. The present book focuses to highlight, how indispensable the Ecogenomics approach is and how the advances in the field of genomics have revolutionized the discipline of ecology.

1.1 INTRODUCTION

Plant ecology in the modern era is a multidisciplinary science with inputs from various fields like molecular biology, biostatistics, biochemistry, biophysics, and other systems of biology. From the early 1990s, plant ecologists have been progressively adopting various techniques and approaches from molecular biology. The use of molecular techniques like the employability of molecular markers to focus on the variability of the DNA of individuals has become an inextricable part of varied sub-disciplines of plant ecology, particularly plant population biology. Traditionally neutral markers such as Simple Sequence Repeats (SSRs) and AFLPs (Amplified Fragment Length Polymorphisms) have been used by ecologists to detect gene flow, inbreeding, genetic drift, and demography. There has been a rapid increase in the knowledge of plant genomics in the last three decades due to advancement in molecular tools, which has revolutionized plant population genetics. Molecular markers are now routinely used for various analytical purposes like molecular taxonomy (Ouborg et al., 1999) phylogenetic relationships (Chase et al., 2005), conservation biology (Slavov et al., 2005); mating system studies (Touzet et al., 2004), the level of inbreeding, (Soltis et al., 2004), finger-printing (Eckstein et al., 2006), genetic, cytogenetic and physical mapping (Eckstein et al., 2006), plant variety protection and marker-assisted selection (MAS) (Magri et al., 2006) in the post-genomic era. However, these neutral markers do not provide an idea about the effect of genetic variation on selection pressure because all these markers based on the repeat regions lie within the non-coding DNA sequences that lack genes. Moreover, due to a lack of knowledge about the function of non-coding DNA, these markers do not allow us to interpret the effect on evolution. With the development and expansion of public genomic databases and the exchange of sequence-related information, it has become possible to develop gene-targeted and functional markers, which are present in or near candidate genes of interest (Andersen and Lübberstedt, 2003). The availability of these markers has further improved the studies related to genetic variability, linkage map construction, and fingerprinting individuals or populations carrying useful genes (Kalendar et al., 2011). These markers have so many applications like selecting parental genotypes, linkage drag elimination during back-crossing, and selecting traits that are not possible through phenotypic characterization (Appleby et al., 2009). As the genome sequences of more organisms are provided, genomic scans are facilitated to differentiate between neutral and non-neutral markers. These advanced Genomic approaches like the next-generation sequencing approach enable ecologists to study if the genes with specific functions are differentially expressed under certain conditions, that is, in response to inbreeding or outbreeding. Next-generation sequencing approaches involving genomic sequences facilitate SNP profiling, which gives an overall estimate of genomic variations. These genomic variations can be used for association studies, which might indicate the genetic mechanisms of plants and their response to the environment. Therefore, research in molecular ecology requires different areas of expertise, and additionally conceptual understanding and integration of mechanistic interactions at different levels is important.

In the ecological arena, this interaction between Ecology and Genomics has led to a new field of research, ecological functional genomics, and this new outlook is significant to search out the solution to the long-enduring queries of evolutionary biology (Timmermans et al., 2008). The ecological genomics approach aims to unravel the genetic basis underlying responses of organisms to the environment. (Feder and Mitchell-Olds, 2003) suggested that in natural environments and populations, this innovative multidisciplinary approach helps to understand the structural and functional aspects of genes with ecological and evolutionary significance and can throw light on the phylogenetic processes and molecular mechanisms that direct phenotypic alterations, elucidate the omics basis of morphological plasticity, and affirm the genomic foundation of selection and the genetic architecture of adaptation within the dynamic and spatially miscellaneous environment. The survival chances of a species in changing environment depend upon how it responds to the modified environment. It involves adaptation response expressed by an organism at morphological, anatomical, physiological, or reproductive levels to cope with the environment. Using genomic tools, it is possible to unravel phenotypic variation's genetic basis and understand how organisms adapt to the climatic variation that ultimately leads to reproductive isolation and speciation. Genomic approaches such as quantitative trait locus (QTL) studies, genome-wide association studies (GWAS), genome scans, and studies on inversions allow for a better understanding of adaptation and validating various theories concerning speciation. By these techniques, it is possible to identify and map ecologically and evolutionarily relevant genes and genomic regions based on their linkage to identifiable markers. Whole-genome polymorphism data from large populations can be used to detect genes related to adaptive phenotypic variations and to scan the footprints of selection across the genome. Transcriptome and proteome help evaluate gene expression changes, separate and identify the proteins to observe the difference in their expression in response to variation in environmental parameters or a specific ecological interaction. Genomic studies on speciation and adaptation in plants specifically focus on adaptation to biotic and abiotic environments, particularly local adaptation, reproductive isolation, gene flow, and the genetic basis of abrupt speciation through interspecific hybridization.

The integration of genomics with ecology, however, requires more than the implementation of a toolbox, since new scientific queries arise with the new technologies and existing questions can be addressed in a way that has not been considered before. For instance, with genomic tools, it has been possible to ascertain profound changes in genome and gene expression in hybrids that make them different from their parents and allow them to inhabit novel habitats inaccessible to their parents. Novel tools like CRISPR/Cas9 system with gene editing ability have facilitated gene expression studies to find the specific role of candidate locus in reproductive isolation. However, to advance in the field of ecological genomics, a further discussion between ecologists and genomics researchers would be required. There are currently two choices for ecologists paying attention to in using genomic tools: one is to use existing model organisms that previously have substantial genomic properties obtainable. This approach is most appropriate to empirical problems that can be solved using model systems. This method is already being widely applied, in particular, in the fields of microbial ecology and plant physiology, and, second, it is using the presently existing methods and applying these to interesting species. The second choice is our priority here, specified that it is the strategy that most ecologists would prefer. Most of the techniques adapted from more conventional genomics research applications for use in ecological genomics need considerable existing resources (Thomas et al., 2004).

1.2 AIMS/APPROACHES OF ECOGENOMICS

For a long, interactions of the organism with the environment and other organisms have been considered of great importance in shaping the future of a community and provide the platform for natural selection in a path to the successful evolution of life forms. The first-ever goal of ecological genomics is to explicate the mechanism of natural selection and adaptive responses. Ecological genomics integrates genomic and ecological approaches to unravel the genetic basis underlying biochemical, physiological, morphological, or behavioral responses of organisms to the environment. Research findings in Ecological and evolutionary discipline have often demonstrated that species, populations, and individuals are adapted to their environment, that is, the fitness of individuals is higher in their home environment than in other environments. The first-ever goal of ecological genomics is to explicate adaptive responses. In particular, plant ecology has a long-standing practice of exploring this adaptation because individual plants have no choice but to adapt to the climate in which they grow. Evolutionary research studied the adaptation of the interaction between evolutionary forces (i.e., more specifically, migration, and drift) and selection, resulting in transformation. The use of genomic tools/techniques now helps to open a black box between the phenotype and the environment to illustrate the adaptive mechanisms while examining phenotypic changes. The second significant goal of Ecogenomics is to evaluate the conceptual mechanism of phenotypic variation and understand how adaptation to local climate ultimately leads to speciation. Genomics can help in the identification of genomic regions and metabolic networks involved in the adaptive response to biotic and abiotic environments (Haak et al., 2014). This objective lines up with the first: while the first objective emphasizes the cause of variation in fitness, the second objective is centered to know the phenotypic variation. These two goals, combined, would provide the foundation for a mechanistic view of evolution. It also investigates the genetic targets for the selection process. Evolutionary influences exist at the genetic level, but the focus of selection is on phenotypes. The introduction of genomic techniques/tools may get used to identify these emerging genes (Lee and Mitchell-Olds, 2006). It is also pertinent to mention here that this ecogenomic approach will also provide the opportunity to understand evolutionary complex phenomena, that is, genetic drawbacks and the origin of phenotypic plasticity (Jackson et al., 2002). Finally, the genomic toolbox offers entirely novel methods of designing environmental studies (Jackson et al., 2002) which not only helps us to investigate diverse ecological and evolutionary processes with unimaginable molecular details but also provides innovative ways of experimental manipulation. Therefore, in Ecogenomics, further developments are expected and that would be the beginning of successful integration of diverse omics approaches, such as transcriptomics and metabolomics, with community ecology (Snoeren et al., 2007). By implementing these technologies, we can conserve biodiversity, minimize environmental pollution, study bioprospecting, improve environmental health, enhance bioremediation, and produces a bio-based economy (DeLong, 2002).

1.3 GENOMIC TOOLS IN SEARCH OF ECOLOGICAL PROBLEMS

Ecological genomics deals with the organism's genomic structure and its interaction with the biotic and abiotic environments (Snoeren et al., 2007). Ecological genomics is the field where the two levels of life, largest and smallest meets that is research study can be performed at different levels, such as proteins to nucleic acids to cells to organisms, and finally at the levels of populations to community up to ecosystems to biospheres. It uses a genomic explanation at the molecular level to describe the relation and interaction between environment and phenotype. Ecogenomics study focuses on the newly formed polymorphic trait for an inquiry at the genome and ecological level (Baldwin et al., 2006; Schranz et al., 2009) and provides knowledge about the genetic mechanism fundamental to the response of an organism to its natural environment. Usually, this query focuses on the level of the organism. In parallel to this ecological genomic analysis, involves elucidating how genomes behave at higher organizational levels, for instance, if there is a "community genome" and we should be able to decipher what role it can play and if so, we should try to understand how it works. We would also like to classify the genes and gene features that function the most in a given ecological relationship while assessing the interaction of organisms with their environments. In recent years, there have been several approaches and techniques in molecular biology used by plant ecologists, especially in plant population biology, and many marker techniques available so for, including AFLP (Amplified fragment length polymorphism), IISSR (Inter simple sequence repeat), and microsatellite-PCR, to find out the variability in coding and non-coding regions. Plant ecologists utilize the top–down approach by applying the metabolomics acquisition methods such as GC-MS, HPLC, LC-MS, and FT-ICR-MS at the plant -diversity-environment interaction level to study the biochemical plasticity and genetic diversity which results in metabolite plasticity. Microarrays, SAGE, and NGS are applied to the monomorphic population to quantify gene expression differences concerning dynamic surroundings. Reverse genetics (gene silencing and overexpression) are applied to manipulate allelic variation to categorize the gene function which is of great importance specifically for the confirmation of gene function after the identification of gene by QTL mapping. Therefore, the plant-ecogenomic approach deals with the impact of surrounding on the genomic constitution of an organism as well as elucidate the interactions between environment and phenotype of model and non-model plant species, using mechanistic and genomic processes. One technique is to explore the role of individual genes whose identification in the sequence indicates that they may be significant for an ecologically relevant pathway or phenotype. For instance, in the natural population of rock pocket mice in Arizona residing on black-colored basalt lava and light-colored rock, Nachman and his companions examined the ecologically significant trait of coat color. Since the genetic regulation of mammalian coat color has been thoroughly researched in the laboratory, multiple candidate loci were the subject of the authors, of which adaptive melanism was found to be associated with mutations in the melanocortin 1 receptor gene (Nachman et al., 2003). Fascinatingly, in many distinct species, this adaptive melanism seems to have arisen spontaneously, and that these variations have an, unlike genetic basis, despite similar phenotypes (Hoekstra and Nachman, 2003). The strength of a candidate gene technique is demonstrated by this and other experiments (Johanson et al., 2000; Stinchcombe et al., 2004). If an approach to the candidate gene is not logical, substitute approaches must be used. Typically, these alternative approaches reflect a "first pass" in determining potentially relevant loci, and further studies must be carried out after this step. To understand environmental disturbances, transcriptional characterization using microarrays can recognize genes whose expression alters and so become candidate genes to be concerned in the response. This is one of the key approaches presently that has been used to classify significant genes in ecological genomics research. Proteomic techniques have always been used to specifically specify proteins that are essential for particular ecological relationships, say two-dimensional gel electrophoresis to isolate proteins from environmental and then categorize these by using mass spectroscopic techniques. However, both methods include functional tests (for instance applying to mutants) to recognize whether or not the genes or proteins are of practical significance. A QTL mapping technique that takes the benefit of regulated crosses and genetic variation that happens naturally is also an effective method. By studying their linkage to polymorphic molecular markers in segregating mapping populations, QTL mapping can provide an efficient tool for localizing the general locations of ecologically and evolutionary important genes. The trust limits on QTL positions, though a common method, usually involve broad chromosome territories and hundreds of genes. Finer scale mapping is essential to further refine the locations of QTLs and is greatly improved if recombination maps and physical maps are incorporated. Another question arises while studying QTLs when organisms are studied *in vitro*, they are grown in controlled conditions, which are isolated from their environmental niche. This facilitates the study of isolated functions and mechanisms, masking the effect of environmental factors. This is evident from an earlier study about the flowering time, one of the widely researched genetic mechanisms, in which mapped QTLs for bolting in controlled environments did not show similar associations in a natural environment. Out of the QTLs mapped in recombinant inbred lines of *Arabidopsis thaliana,* some QTLs were not detected in a natural environment and others had been redetected in a specific set of environments. Additionally, some novel QTLs were detected in natural conditions (Weinig et al., 2002). Such findings of a best-studied trait illustrate the complexity of environmental dependence. Next-generation sequencing technology provides a simple means to assess the sequence diversity in an organism. In the present context with the evolution of the sequencing approach and reduction in the cost of sequencing, the future of studying the genome, transcriptome, and epigenome of any non-model species looks promising (Metzker, 2010). Within the last few years, the cost of sequencing has been drastically reduced with the advancement of technology. Many model organisms have been successfully sequenced and genome assemblies have been properly annotated. Transcriptomes and genomes sequenced from non-model species can be aligned on a properly annotated reference genome of the model organism. Although to eliminate the necessity of a reference assembly, many methods have been developed to perform *de novo* analyses (Forouzan et al., 2017). Next-generation sequencing approaches involving genomic sequences facilitate SNP profiling, which gives an overall estimate of genomic variations including diversity studies; molecular taxonomy; phylogenetic relationships; conservation biology; mating systems; the level of inbreeding; fingerprinting; genetic, cytogenetic, and physical mapping; diagnostics; plant variety protection and MAS in the post-genomic era. These genomic variations can be used for association studies, which might indicate the genetic mechanisms of plants and their response to the environment.

However, all of the methods that have been mentioned above profits significantly from the currently known genomic tools only in certain species. In certain ecological experiments, the chosen species do not have these resources available. So, what concessions should be made, if any? Will it be possible to research the ecology of selected species that may not be much informative or does it improve the genomic abilities of more ecologically relevant taxa? Both methods have provided impressive findings at this point (Roberts and Feder, 2000), (Weinig et al., 2002). Although transferring the genomic information from model organisms to non-model ones solves some questions. Expansion of sequencing approach to these non-model species amenable to study gives the field more strength. Whole genome sequencing projects running in different research centers have sequenced 1000 plant reference genomes which have enriched the database (Wang et al., 2020). Furthermore, functional tests of genetic outcomes of the studies are also required for which new methodologies have to be developed. Sometimes genomic scans might fail to map local adaptation genes because of the coupling of intrinsic pre- or post-zygotic incompatibilities that create endogenous barriers to gene flow and exogenous barriers like environmental constraints (Bierne et al., 2011). The studies on genomic structure of populations of *Bochera stricta* and others have used a collective approach (Song et al., 2006), and the results have proven to be exceptionally successful. Other compromising approach includes the cross-hybridization of RNAs from one species to chips of genes produced for other similar species (Renn, 2004). Finally, as genetic model systems, many ecologically relevant species are currently being created (Gewin, 2005). In particular, the genome sequences in the case of *Daphnia pulex* (Colbourne et al., 2005), *Gasterosteus aculeatus* (Peichel et al., 2001; Colosimo et al., 2005; Peichel, 2005), and *Populus trichocarpa* (Busov et al., 2005; DiFazio, 2005) have been identified and related genetic tools have been developed. While this model versus non-model issue will remain to be addressed in the community of ecological genomics, we ultimately predict no single approach to ecological genomics to be the solution. Further, the separate or collective use of model and non-model systems, either in the same research program or in different programs together, will continue to produce substantial results. It remains a more complex and uphill task to organize how genomes operate at higher levels of organization. Metagenomic studies of microbial communities are the best and most compelling achievements to date in this field (Handelsman, 2004). Metagenomic research includes isolating DNA from environmental isolates, cloning it, sequencing the clones and organizing the representative genomes into large or small insert libraries. By using taxon-oriented markers like 16S rRNA genes, large inserts help in the identification of the sequences that are phylogenetically significant (Allen and Banfield, 2005). Metagenomics may also show the degree to which species inside microbial communities' interplay as consortia, offering a crucial role, along with recognition of novel species and pathways. Analysis in the case of acid mine drainage biofilms offers an outstanding example (Tyson et al., 2004). Nowadays ample relevant studies are carried out involving population microarrays and proteomics and are important for the community to evaluate the gene functions (Ram et al., 2005). Further, the use of functional microarrays of genes that measure the existence of genes involved in the cycling of carbon and nitrogen, in particular, can also help define essential community roles (Schadt et al., 2004). Thus, all the ecogenomic approaches described above differ in their resolving power to detect genetic differences, the type of data they generate, and their applicability to particular taxonomic levels. Along with this, further studies need to be done to understand complex communities as found in the case of soil that will be the future of community genomics. This book presents a wide description of the advancement of different molecular techniques that can be utilized for the analysis of plant population genetics and the different challenges and opportunities.

1.4 CHALLENGES FOR ECOGENOMICS

The field of Ecogenomics has great potential to answer several ecologically relevant questions still certain challenges act as a remarkable hurdle towards the usefulness of this branch of science. In natural ecosystems, the first challenge is to expand the eco-genomic approaches to non-model organisms to offer a broad range of ecological conditions and life histories. There has been a lot of omics work done on the model and non-model crop species which aims to widen the understanding of subcellular functioning, for example, the evolution of plant herbivory defense strategy and their gene expression pattern exploration. Molecular tools used in this context, for model plant species can also be applied upon more closely related non-model species for eco-genomic studies (Mitchell-Olds, 2001). Eco-genomic tools in *Arabidopsis* are applied to study the genetic basis of plant defense strategies for its relatives *Brassica, Boechera,* and other members of the Brassicaceae (Broekgaarden et al., 2007). But for a detailed understanding of the evolution and existence of plant defense traits, it is necessary to apply Ecogenomics on other species belonging to different taxas. But the genomic data are not available for many ecologically important species that's why it is still a problem that has to be resolved. Further, the integration of the genomics approach in non-model species becomes more difficult (Tagu et al., 2014). There are many additional reasons for it like large genome size and shallow knowledge about the genetic makeup. Polyploidy in plants makes genome sequencing intensive and therefore restricts the rapid development of knowledge about genome-wide sequences of non-model species (Pinu et al., 2019). Though, this scenario will change with the advent of next-generation sequencing. So, we can contemplate this as a crucial era for ecological omics research (Sardans and Peñuelas, 2013).

Another important challenge for the ecogenomic approach is to integrate ecology and evolutionary biology's experimental languages into molecular biology. Ecology and molecular biology follow separate empirical procedures. Molecular biologists prefer to work in more or less homogenous regulated environments trying to eradicate as much heterogeneity as possible (Ouborg and Vriezen, 2007). In Contrast, Ecologists consider environmental disparity and genotypic variation in the study then accordingly design the experiment. A study on the impact of high CO2 concentrations on the physiological functioning of *Populus* trees provides an example of the disparity among ecological and genomic research designs. The difference in morphological and physiological responses to CO2 elevation was analyzed by using three CO2 stages, descendants from three separate crosses, and a total of 324 biological replicates were used (Liu et al., 2006). The impact of higher CO2 concentration on the expression of *Populus* genes was analyzed by using two levels of CO2, one genetic clone, and 12 biological replicates in total (Taylor et al., 2005). The difficulty lies in linking both results as they show variability in magnitude. Therefore, to incorporate both the subcellular realm and natural environment it is essential to design the experiments which integrate the data from both fields in a complete sense (Taylor et al., 2005).

The development of functional tests to validate the outcome of large sequencing data analyses for a large number of individuals is a challenging task. In an ecological design, using high-throughput instruments, such as microarrays, will produce the expression data for thousands of genes for each of hundreds of individuals. The large amount of data produced will be subjected to various sources of variance including methodological, factorial, and error sources. Improving the quality and quantity of the data cost-effectively is the next challenge to deal with. Data storage and compaction algorithms compatible particularly with DNA sequence information are needed to be worked upon. An international collaboration, both at the level of proficiency as well as research infrastructure, is also needed. This includes an open and outreaching approach to collaboration and authorship choices.

1.5 CONCLUSION AND FUTURE PROSPECTS

Ecogenomics is the study of the effect of ecological constraints on the structure and function of genomes. This vast area of science is especially devoted to gain an understanding of the relationship between an organism and its environment. It not only deals with the relationship of the individual to its surroundings but also covers the way of adjustments and mechanisms involved during the interaction of the whole ecosystem at the community level. Thus, this field is recognized as an emerging area for a better understanding of links between molecular techniques and ecology-related genomic obstacles like mutations and epigenetic alterations (DNA methylation, acetylation, phosphorylation, ubiquitination, etc.). There are several applications of plant-eco-genomic approaches, among them studies to address a variety of ecological phenomena, including plant reproductive biology (Nasrallah et al., 2002; Bastow et al., 2004), flowering time (Koornneef et al., 2004; Stinchcombe et al., 2004), herbivory (Voelckel and Baldwin, 2004; Ralph et al., 2006; Snoeren et al., 2007), plant pathogenic interactions (Allen et al., 2004), drought resistance (Knight et al., 2006), heavy metal tolerance (Rigola et al., 2006) and speciation (Hegarty et al., 2005) have gained more attention. Despite a plethora of opportunities and advantages, studies related to ecological genomics harbor seldom hindrances and limitations. Major obstacles are: problems in the maintenance of large samples and sequencing of their genome, available literature pertaining only model plant species like *Arabidopsis thaliana, Medicago sp.* and a dearth of tools to analyze Ecogenomics data. Thus, there should be more concern towards non-model but economically, taxonomically, and evolutionary important plants and inventions of more techniques and tools to justify the reliability of molecular data related to ecological behavior of tester plant species.

KEYWORDS

- **ecogenomics**
- **multidisciplinary**
- **molecular techniques**
- **population biology**
- **polymorphism**

REFERENCES

Allen, E. E.; Banfield, J. F. Community Genomics in Microbial Ecology and Evolution. *Nat. Rev. Microb.* **2005,** *3* (6), 489–498. https://doi.org/10.1038/nrmicro1157

Allen, R. L.; Bittner-Eddy, P. D.; Grenville-Briggs, L. J.; Meitz, J. C.; Rehmany, A. P.; Rose, L. E.; Beynon, J. L. Host-Parasite Coevolutionary Conflict between Arabidopsis and Downy Mildew. *Science* **2004,** *306* (5703), 1957–1960. https://doi.org/10.1126/science.1104022

Andersen, J. R.; Lübberstedt, T. Functional Markers in Plants. *Trends Plant Sci.* **2003,** *8* (11), 554–560. https://doi.org/10.1016/j.tplants.2003.09.010

Appleby, N.; Edwards, D.; Batley, J. New Technologies for Ultra-High Throughput Genotyping in Plants. *Methods Mol. Biol.* **2009,** *513*, 19–39. https://doi.org/10.1007/978-1-59745-427-8_2

Baldwin, C.; Hienerth, C.; Hippel, E. How User Innovations Become Commercial Products: A Theoretical Investigation and Case Study. *Res. Policy* **2006,** *35*, 1291–1313. https://doi.org/10.1016/j.respol.2006.04.012

Bastow, R.; Mylne, J. S.; Lister, C.; Lippman, Z.; Martienssen, R. A.; Dean, C. Vernalization Requires Epigenetic Silencing of FLC by Histone Methylation. *Nature* **2004,** *427* (6970), 164–167. https://doi.org/10.1038/nature02269

Bierne, N.; Welch, J.; Loire, E.; Bonhomme, F.; David, P. The Coupling Hypothesis: Why Genome Scans May Fail to Map Local Adaptation Genes. *Mol. Ecol.* **2011,** *20* (10), 2044–2072. https://doi.org/10.1111/j.1365-294X.2011.05080.x

Broekgaarden, C.; Poelman, E. H.; Steenhuis, G.; Voorrips, R. E.; Dicke, M.; Vosman, B. Genotypic Variation in Genome-Wide Transcription Profiles Induced by Insect Feeding: Brassica Oleracea—Pieris Rapae Interactions. *BMC Genom.* **2007,** *8*, 239. https://doi.org/10.1186/1471-2164-8-239

Busov, V. B.; Brunner, A. M.; Meilan, R.; Filichkin, S.; Ganio, L.; Gandhi, S.; Strauss, S. H. Genetic Transformation: A Powerful Tool for Dissection of Adaptive Traits in Trees. *New Phytol.* **2005,** *167* (1), 9–18. https://doi.org/10.1111/j.1469-8137.2005.01412.x

Chase, M. W.; Salamin, N.; Wilkinson, M.; Dunwell, J. M.; Kesanakurthi, R. P.; Haider, N.; Savolainen, V. Land Plants and DNA Barcodes: Short-Term and Long-Term Goals. *Philos. Trans. R. Soc. Lond. B Biol. Sci.* **2005,** *360* (1462), 1889–1895. https://doi.org/10.1098/rstb.2005.1720

Colbourne, J. K.; Singan, V. R.; Gilbert, D. G. wFleaBase: The Daphnia Genome Database. *BMC Bioinfo.* **2005,** *6*, 45. https://doi.org/10.1186/1471-2105-6-45

Colosimo, P. F.; Hosemann, K. E.; Balabhadra, S.; Villarreal, G.; Jr.; Dickson, M.; Grimwood, J.; Schmutz, J.; Myers, R. M.; Schluter, D.; Kingsley, D. M. Widespread Parallel Evolution in Sticklebacks by Repeated Fixation of Ectodysplasin Alleles. *Science* **2005,** *307* (5717), 1928–1933. https://doi.org/10.1126/science.1107239

DeLong, E. F. Microbial Population Genomics and Ecology. *Curr. Opin. Microb.* **2002,** *5* (5), 520–524. https://doi.org/10.1016/s1369-5274(02)00353-3

DiFazio, S. P. A Pioneer Perspective on Adaptation. *New Phytol.* **2005,** *165* (3), 661–664. https://doi.org/https://doi.org/10.1111/j.1469-8137.2005.01334.x

Eckstein, R. L.; O'Neill R, A.; Danihelka, J.; Otte, A.; Köhler, W. Genetic Structure among and within Peripheral and Central Populations of Three Endangered Floodplain Violets. *Mol. Ecol.* **2006,** *15* (9), 2367–2379. https://doi.org/10.1111/j.1365-294X.2006.02944.x

Feder, M. E.; Mitchell-Olds, T. Evolutionary and Ecological Functional Genomics. *Nat. Rev. Genet.* **2003,** *4* (8), 651–657. https://doi.org/10.1038/nrg1128

Forouzan, E.; Maleki, M. S. M.; Karkhane, A. A.; Yakhchali, B. Evaluation of Nine Popular de novo Assemblers in Microbial Genome Assembly. *J. Microbiol. Methods* **2017,** *143*, 32–37. https://doi.org/https://doi.org/10.1016/j.mimet.2017.09.008

Gewin, V. Functional Genomics Thickens the Biological Plot. *PLoS Biol* **2005,** *3* (6), e219. https://doi.org/10.1371/journal.pbio.0030219

Haak, D. C.; Kostyun, J. L.; Moyle, L. C. Merging Ecology and Genomics to Dissect Diversity in Wild Tomatoes and Their Relatives. *Adv. Exp. Med. Biol.* **2014,** *781*, 273–298. https://doi.org/10.1007/978-94-007-7347-9_14

Handelsman, J. Metagenomics: Application of Genomics to Uncultured Microorganisms. *Microbiol. Mol. Biol. Rev.* **2004,** *68* (4), 669–685. https://doi.org/10.1128/mmbr.68.4.669-685.2004

Hegarty, M. J.; Jones, J. M.; Wilson, I. D.; Barker, G. L.; Coghill, J. A.; Sanchez-Baracaldo, P.; Liu, G.; Buggs, R. J.; Abbott, R. J.; Edwards, K. J.; Hiscock, S. J. Development of Anonymous cDNA Microarrays to Study Changes to the Senecio Floral

Transcriptome during Hybrid Speciation. *Mol. Ecol.* **2005,** *14* (8), 2493–2510. https://doi.org/10.1111/j.1365-294x.2005.02608.x
Hoekstra, H. E.; Nachman, M. W. Different Genes Underlie Adaptive Melanism in Different Populations of Rock Pocket Mice. *Mol. Ecol.* **2003,** *12* (5), 1185–1194. https://doi.org/10.1046/j.1365-294x.2003.01788.x
Jackson, R. B.; Linder, C. R.; Lynch, M.; Purugganan, M.; Somerville, S.; Thayer, S. S. Linking Molecular Insight and Ecological Research. *Trends Ecol. Evol.* **2002,** *17* (9), 409–414. https://doi.org/https://doi.org/10.1016/S0169-5347(02)02571-5
Johanson, U.; West, J.; Lister, C.; Michaels, S.; Amasino, R.; Dean, C. Molecular Analysis of FRIGIDA, a Major Determinant of Natural Variation in Arabidopsis Flowering Time. *Science* **2000,** *290* (5490), 344–347. https://doi.org/10.1126/science.290.5490.344
Kalendar, R.; Flavell, A. J.; Ellis, T. H.; Sjakste, T.; Moisy, C.; Schulman, A. H. Analysis of Plant Diversity with Retrotransposon-based Molecular Markers. *Heredity (Edinb)* **2011,** *106* (4), 520–530. https://doi.org/10.1038/hdy.2010.93
Knight, C. A.; Vogel, H.; Kroymann, J.; Shumate, A.; Witsenboer, H.; Mitchell-Olds, T. Expression Profiling and Local Adaptation of *Boechera holboellii* Populations for Water Use Efficiency across a Naturally Occurring Water Stress Gradient. *Mol. Ecol.* **2006,** *15* (5), 1229–1237. https://doi.org/10.1111/j.1365-294X.2006.02818.x
Koornneef, M.; Alonso-Blanco, C.; Vreugdenhil, D. Naturally Occurring Genetic Variation in *Arabidopsis thaliana*. *Annu. Rev. Plant Biol.* **2004,** *55*, 141–172. https://doi.org/10.1146/annurev.arplant.55.031903.141605
Lee, C. E.; Mitchell-Olds, T. Preface to the Special Issue: Ecological and Evolutionary Genomics of Populations in Nature. *Mol. Ecol.* **2006,** *15* (5), 1193–1196. https://doi.org/10.1111/j.1365-294X.2006.02945.x
Liu, N.; Dang, Q. L.; Parker, W. H. Genetic Variation of Populus Tremuloides in Ecophysiological Responses to CO2 Elevation. *Can. J. Bot. J. Can. de botanique* **2006,** *84* (2), 294–302. https://doi.org/10.1139/b05-171
Magri, D.; Vendramin, G. G.; Comps, B.; Dupanloup, I.; Geburek, T.; Gömöry, D.; Latałowa, M.; Litt, T.; Paule, L.; Roure, J. M.; Tantau, I.; van der Knaap, W. O.; Petit, R. J.; de Beaulieu, J. L. A New Scenario for the Quaternary History of European Beech Populations: Palaeobotanical Evidence and Genetic Consequences. *New Phytol* **2006,** *171* (1), 199–221. https://doi.org/10.1111/j.1469-8137.2006.01740.x
Metzker, M. L. Sequencing Technologies—the Next Generation. *Nat. Rev. Genet.* **2010,** *11* (1), 31–46. https://doi.org/10.1038/nrg2626
Mitchell-Olds, T. *Arabidopsis thaliana* and Its Wild Relatives: A Model System for Ecology and Evolution. *Trends Ecol. Evol.* **2001,** *16* (12), 693–700. https://doi.org/https://doi.org/10.1016/S0169-5347(01)02291-1
Nachman, M. W.; Hoekstra, H. E.; D'Agostino, S. L. The Genetic Basis of Adaptive Melanism in Pocket Mice. *Proc. Natl. Acad. Sci. USA* **2003,** *100* (9), 5268–5273. https://doi.org/10.1073/pnas.0431157100
Nasrallah, M. E.; Liu, P.; Nasrallah, J. B. Generation of Self-incompatible *Arabidopsis thaliana* by Transfer of Two S Locus Genes from A. Lyrata. *Science* **2002,** *297* (5579), 247–249. https://doi.org/10.1126/science.1072205
Ouborg, N. J.; Piquot, Y.; Van Groenendael, J. M. Population Genetics, Molecular Markers and the Study of Dispersal in Plants. *J. Ecol.* **1999,** *87* (4), 551–568. https://doi.org/https://doi.org/10.1046/j.1365-2745.1999.00389.x

Ouborg, N. J.; Vriezen, W. H. An Ecologist's Guide to Ecogenomics. *J. Ecol.* **2007,** *95* (1), 8–16. https://doi.org/https://doi.org/10.1111/j.1365-2745.2006.01197.x

Peichel, C. L. Fishing for the Secrets of Vertebrate Evolution in Threespine Sticklebacks. *Dev. Dyn.* **2005,** *234* (4), 815–823. https://doi.org/10.1002/dvdy.20564

Peichel, C. L.; Nereng, K. S.; Ohgi, K. A.; Cole, B. L.; Colosimo, P. F.; Buerkle, C. A.; Schluter, D.; Kingsley, D. M. The Genetic Architecture of Divergence between Threespine Stickleback Species. *Nature* **2001,** *414* (6866), 901–905. https://doi.org/10.1038/414901a

Pinu, F. R.; Beale, D. J.; Paten, A. M.; Kouremenos, K.; Swarup, S.; Schirra, H. J.; Wishart, D. Systems Biology and Multi-Omics Integration: Viewpoints from the Metabolomics Research Community. *Metabolites* **2019,** *9* (4), 76. https://www.mdpi.com/2218-1989/9/4/76

Ralph, S. G.; Yueh, H.; Friedmann, M.; Aeschliman, D.; Zeznik, J. A.; Nelson, C. C.; Butterfield, Y. S.; Kirkpatrick, R.; Liu, J.; Jones, S. J.; Marra, M. A.; Douglas, C. J.; Ritland, K.; Bohlmann, J. Conifer Defence against Insects: Microarray Gene Expression Profiling of *Sitka spruce* (*Picea sitchensis*) Induced by Mechanical Wounding or Feeding by Spruce Budworms (*Choristoneura occidentalis*) or White Pine Weevils (*Pissodes strobi*) Reveals Large-Scale Changes of the Host Transcriptome. *Plant Cell Environ.* **2006,** *29* (8), 1545–1570. https://doi.org/10.1111/j.1365-3040.2006.01532.x

Ram, R.; Verberkmoes, N.; Thelen, M.; Tyson, G.; Baker, B.; Blake, R.; Shah, M.; Hettich, R.; Banfield, J. Community Proteomics of a Natural Microbial Biofilm. *Science* **2005,** *308*, 1915–1920. https://doi.org/10.1126/science.+1109070

Renn, O. Perception of Risks. *Toxicol. Lett.* **2004,** *149* (1), 405–413. https://doi.org/https://doi.org/10.1016/j.toxlet.2003.12.051

Rigola, D.; Fiers, M.; Vurro, E.; Aarts, M. G. The Heavy Metal Hyperaccumulator Thlaspi Caerulescens Expresses Many Species-Specific Genes, as Identified by Comparative Expressed Sequence Tag Analysis. *New Phytol.* **2006,** *170* (4), 753–765. https://doi.org/10.1111/j.1469-8137.2006.01714.x

Roberts, S. P.; Feder, M. E. Changing Fitness Consequences of hsp70 Copy Number in Transgenic *Drosophila larvae* Undergoing Natural Thermal Stress. *Funct. Ecol.* **2000,** *14* (3), 353–357. https://doi.org/https://doi.org/10.1046/j.1365-2435.2000.00429.x

Sardans, J.; Peñuelas, J. Tree Growth Changes with Climate and Forest Type Are Associated with Relative Allocation of Nutrients, Especially Phosphorus, to Leaves and Wood. *Global Ecol. Biogeo.* **2013,** *22* (4), 494–507. https://doi.org/https://doi.org/10.1111/geb.12015

Schadt, C.; Liebich, J.; Chong, S.; Gentry, T.; He, Z.; Pan, H.; Zhou, J. Design and Use of Functional Gene Microarrays (FGAs) for the Characterization of Microbial Communities. **2004,** *34*, 331–368.

Schranz, M. E.; Manzaneda, A. J.; Windsor, A. J.; Clauss, M. J.; Mitchell-Olds, T. Ecological Genomics of *Boechera stricta*: Identification of a QTL Controlling the Allocation of Methionine- vs Branched-Chain Amino Acid-Derived Glucosinolates and Levels of Insect Herbivory. *Heredity (Edinb)* **2009,** *102* (5), 465–474. https://doi.org/10.1038/hdy.2009.12

Slavov, G. T.; Howe, G. T.; Gyaourova, A. V.; Birkes, D. S.; Adams, W. T. Estimating Pollen Flow Using SSR Markers and Paternity Exclusion: Accounting for

Mistyping. *Mol. Ecol.* **2005,** *14* (10), 3109–3121. https://doi.org/https://doi.org/10.1111/j.1365-294x.2005.02620.x

Snoeren, T.; Jong, P.; Dicke, M. Ecogenomic Approach to the Role of Herbivore-Induced Plant Volatiles in Community Ecology. *J. Ecol.* **2007,** *95* (1), 17–26. https://doi.org/10.1111/j.1365-2745.2006.01183.x

Soltis, D. E.; Albert, V. A.; Savolainen, V.; Hilu, K.; Qiu, Y.-L.; Chase, M. W.; Farris, J. S.; Stefanović, S.; Rice, D. W.; Palmer, J. D.; Soltis, P. S. Genome-Scale Data, Angiosperm Relationships, and "Ending Incongruence": A Cautionary Tale in Phylogenetics. *Trends Plant Sci.* **2004,** *9* (10), 477–483. https://doi.org/10.1016/j.tplants.2004.08.008

Song, B. H.; Clauss, M. J.; Pepper, A.; Mitchell-Olds, T. Geographic Patterns of Microsatellite Variation in *Boechera stricta*, a Close Relative of Arabidopsis. *Mol. Ecol.* **2006,** *15* (2), 357–369. https://doi.org/10.1111/j.1365-294X.2005.02817.x

Stinchcombe, J. R.; Weinig, C.; Ungerer, M.; Olsen, K. M.; Mays, C.; Halldorsdottir, S. S.; Purugganan, M. D.; Schmitt, J. A Latitudinal Cline in Flowering Time in *Arabidopsis thaliana* Modulated by the Flowering Time Gene FRIGIDA. *Proc. Natl. Acad. Sci. USA* **2004,** *101* (13), 4712–4717. https://doi.org/10.1073/pnas.0306401101

Tagu, D.; Colbourne, J. K.; Nègre, N. Genomic Data Integration for Ecological and Evolutionary Traits in Non-Model Organisms. *BMC Genom.* **2014,** *15* (1), 490. https://doi.org/10.1186/1471-2164-15-490

Taylor, G.; Street, N. R.; Tricker, P. J.; Sjödin, A.; Graham, L.; Skogström, O.; Calfapietra, C.; Scarascia-Mugnozza, G.; Jansson, S. The Transcriptome of Populus in Elevated CO2. *New Phytol.* **2005,** *167* (1), 143–154. https://doi.org/10.1111/j.1469-8137.2005.01450.x

Thomas, C. D.; Cameron, A.; Green, R. E.; Bakkenes, M.; Beaumont, L. J.; Collingham, Y. C.; Erasmus, B. F.; De Siqueira, M. F.; Grainger, A.; Hannah, L.; Hughes, L.; Huntley, B.; Van Jaarsveld, A. S.; Midgley, G. F.; Miles, L.; Ortega-Huerta, M. A.; Peterson, A. T.; Phillips, O. L.; Williams, S. E. Extinction Risk from Climate Change. *Nature* **2004,** *427* (6970), 145–148. https://doi.org/10.1038/nature02121

Timmermans, M. J.; Roelofs, D.; Mariën, J.; van Straalen, N. M. Revealing Pancrustacean Relationships: Phylogenetic Analysis of Ribosomal Protein Genes Places Collembola (springtails) in a Monophyletic Hexapoda and Reinforces the Discrepancy between Mitochondrial and Nuclear DNA Markers. *BMC Evol. Biol.* **2008,** *8*, 83. https://doi.org/10.1186/1471-2148-8-83

Touzet, P.; Hueber, N.; Bürkholz, A.; Barnes, S.; Cuguen, J. Genetic Analysis of Male Fertility Restoration in Wild Cytoplasmic Male Sterility G of Beet. *Theor. Appl. Genet.* **2004,** *109* (1), 240–247. https://doi.org/10.1007/s00122-004-1627-7

Tyson, G. W.; Chapman, J.; Hugenholtz, P.; Allen, E. E.; Ram, R. J.; Richardson, P. M.; Solovyev, V. V.; Rubin, E. M.; Rokhsar, D. S.; Banfield, J. F. Community Structure and Metabolism through Reconstruction of Microbial Genomes from the Environment. *Nature* **2004,** *428* (6978), 37–43. https://doi.org/10.1038/nature02340

Voelckel, C.; Baldwin, I. T. Generalist and Specialist Lepidopteran Larvae Elicit Different Transcriptional Responses in *Nicotiana attenuata*, Which Correlate with Larval FAC Profiles. *Ecol. Lett.* **2004,** *7* (9), 770–775. https://doi.org/https://doi.org/10.1111/j.1461-0248.2004.00633.x

Wang, T.; Hu, J.; Ma, X.; Li, C.; Yang, Q.; Feng, S.; Li, M.; Li, N.; Song, X. Identification, Evolution and Expression Analyses of Whole Genome-wide TLP Gene Family in *Brassica napus*. *BMC Genom.* **2020,** *21* (1), 264. https://doi.org/10.1186/s12864-020-6678-x

Weinig, C.; Ungerer, M. C.; Dorn, L. A.; Kane, N. C.; Toyonaga, Y.; Halldorsdottir, S. S.; Mackay, T. F.; Purugganan, M. D.; Schmitt, J. Novel Loci Control Variation in Reproductive Timing in *Arabidopsis thaliana* in Natural Environments. *Genetics* **2002,** *162* (4), 1875–1884.

CHAPTER 2

Advancement in Molecular Tools of Plant Population Genetics

DIVYA RAJ PRASAD[1,*], TARUN MINOCHA[2], BIPIN MAURYA[3], and PEERZADA YASIR YOUSUF[4]

[1]*Department of Kayachikitsa, Institute of Medical Sciences, Banaras Hindu University, Varanasi 221005, India*

[2]*Department of Zoology, Institute of Science, Banaras Hindu University, Varanasi 221005, India*

[3]*Department of Botany, Institute of Science, Banaras Hindu University, Varanasi 221005, India*

[4]*Departmentof Botany, Government Degree College Pulwama, Jammu and Kashmir 192301, India*

[*]*Corresponding author. Email: divyarajbhu@gmail.com*

ABSTRACT

The living world comprises an incredible diversity of flora that encompasses an endless variety of plant species with aesthetic and medicinal values. To evaluate the genetic diversity of plants concerning analyzing the specific genes and their mode of gene action, construction of genetic and physical maps, and for assisting the progress of gene transfer technologies as techniques a large array of molecular markers are employed in the plant kingdom. Showing a crucial role in the studies of species evolution, phylogeny and conservation molecular techniques are applied to broaden

Plant Ecogenomics. Peerzada Arshid Shabir, PhD, Peerzada Yasir Yousuf, PhD & Khalid Rehman Hakeem, PhD (Eds.)

our understanding of the distribution and amount of genetic diversity not just within the species but also between the species. This chapter is the epitome of the new advances made in the molecular marker techniques along with their applications, advantages, and limitations and serves as a concrete platform forming a base for the study in the population genetics for future research directions.

2.1 INTRODUCTION

As we know that plant genomes are highly diverse, and these genomic variations can help us understand the biological phenomena within different plant populations. Gene flow, population density, mating system, and evolutionary framework are some of the key factors that determine variations in the plant population. Diverse agronomic and morphological parameters were successfully used to evaluate genomic diversity and other important molecular characteristics. There has been a rapid increase in the knowledge of plant genomes in the last three decades due to advancement in molecular tools, which has revolutionized the plant population genetics.

Population genomics is one of the facets of life sciences in which DNA sequences of the population are compared with each other. It observes patterns in the genome among the population and inside it so that further these patterns can conclude about genome and population. Population genomics enhances the understanding of phylogenetic history, conservation, ecology, agriculture, forestry, demography of population, and human health in a very accurate manner. "Genetic structure" is defined as any pattern in the genetic makeup of individuals within the population. It provides information about an individual to be concluded from another individual of the same population. Genetic structure talks about evolutionary history and population potential to evolve that's why it is the fourth most step in the study of plant population genetics. Evolution occurs due to variation in fitness among individuals in the environment caused by natural selection. This book chapter serves the purpose of reviewing the advancement in molecular tools and techniques in the study of plant population genetics. Many options are there for studies of plant population in the field of molecular techniques. Distinct forms of molecular tools and markers are being used for different types of plants in the Plant Kingdom like forest trees, shrubs, herbs, climbers, creepers, and fungi. This chapter also reveals some basic and important questions in plant population genomics and molecular evolution of plant kingdom and their cultivated relatives. We review how whole-genome sequencing of intact population increased the speed of invention of the mechanisms of plant genome that further helps evolution the evolutionary forces to frame genetic diversity within and among species and culminate recent exceptional examples from the literature. In this whole chapter, we focus mostly on the population size of plant genome evolution and confer about the reason plant genomes regularly show limited evidence of adaptation and display features that are inconsistent with simple models of linked selection.

2.2 HISTORY

In the year 1980s showed changes in the fate of plant genetics and breeding with the development of molecular marker technology. To define the role of molecular tools and markers, myriad of research was executed in the last three decades. Development and advancement in sequencing technologies help in the improvement of different types of crops. This review adds value to the previous knowledge about molecular markers and their application in plant breeding. The advancement made in genetics, genome editing, molecular plant breeding, and genome selection played a crucial role in the extensive understanding of molecular markers and gives a broad approach to crop diversity, and hugely supplemented breeding plans and preparations. Association mapping based on next-generation sequencing technologies and genotyping-by-sequencing promotes the identification of novel genetic markers for the composite and unorganized population. This chapter comprises various aspects of molecular markers and their functions in plant breeding and population and their growth in genome selection and genome editing.

The molecular markers of forest trees using the plant population genetics have to match certain conditions like co-dominance of alleles, ease, and speed of genotyping, reproducibility over time, and space, the possibility of easily exchanging marker information, and high information content. Not even a single marker system available in the market fully matches all these above requirements. After seeing all the pros and cons of each of the marker systems, the use of simple sequence repeats (SSRs or microsatellites) was paid attention to because they are co-dominant, profoundly informative, reproducible, and easily exchangeable.

In the category of shrubs, the rose is the best example which is a woody perennial flowering plant and is mostly used as an influential ornamental plant around the globe. Besides the monetary importance, researchers are trying to interpret the important characters inherited in the roses. For the selection of molecular tools used in the research of rose and its breeding, first, analysis of morphological characters and resistance to two important pathogens i.e., powdery mildew, and black spot, which molecularly and genetically is mandatory. Mapping of the morphological traits like single genes and QTLs (quantitative trait loci) to molecular markers was done. Mapping of a resistance gene for each of the diseases, plant markers like powdery conferring high-density mapping, resistance against black spot, and isolation of BAC-Clones with closely linked markers resulted in recognition of a contig carrying several disease resistance analogous sequences which spanned the locus. The current investigation occurs on the structure of this locus and the understanding of Rdr1 among the candidate genes. Plans for the marker-assisted choice against the genetic background of wild donor species in introgression programs can be generated in near future. The efficiency of wild donor germplasm can be lowered more adequately from molecular markers than a traditional method by backcrossing within two generations.

In the category of trees, cashew is a perennial evergreen tree grown for its edible fruits. The genetic improvement of the traditional method in cashew is gradual and fickle due to the large severity of heterozygosity, the huge size of the mature plant, the long juvenile phase, and lack of juvenile-mature traits correlations. The complication of breeding increased due to yield and its agronomic traits. Nowadays, next-generation sequencing (NGS) and high-throughput genotyping technologies have accelerated the advancement of genomic tools and resources for orphan crops like cashew nuts.

2.3 MOLECULAR BIOLOGY TOOLS

Drastic progress in molecular biology tools (MBTs) has modified the custom of species identification lately. This advancement not only has provided scientists a vivid range of molecular techniques for accessible and solid identification of plant species but also is an aid in the conservation of plant genetic resources. MBTs were developed by scientists to detect cells, RNA, DNA, and proteins for their research program. Most of the advanced techniques have been profitably used to study distribution and degree of alteration in species gene pools and to solve the typical evolutionary and taxonomic problems (Fig. 2.1).

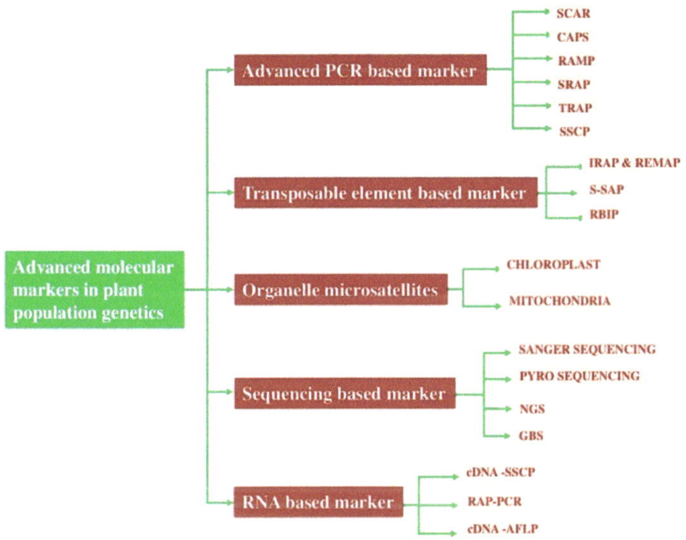

FIGURE 2.1 Schematic representation of advanced molecular tools used in plant population genetics.

Molecular tools can be categorized into three basic categories and genetic variations within and between plant populations can be analyzed by three techniques:

(a) Morphological markers
(b) Biochemical markers
(c) Molecular or DNA marker

2.3.1 *MORPHOLOGICAL MARKERS*

Markers based on the morphology of plants such as plant length, flower color, seed size, seed color, etc. are termed morphological markers. It is based on plant morphology so it is inexpensive but time-consuming and need a broad area for crop trials. Morphological markers are highly dependable on environmental factors and change in the environment produces phenotypic plasticity in the plant population.

2.3.1.1 ADVANTAGES OF MORPHOLOGICAL MARKERS

- Morphological markers are readily available.
- These markers are successfully used in the breeding programs for different crops.

2.3.1.2 DISADVANTAGES OF MORPHOLOGICAL MARKERS

- Economic traits including yield and quality are not associated with these types of markers.
- The growth and development of plants are related to these markers showing undesirable effects on them.
- Though morphological mutants occur rarely and are highly detrimental and lethal, morphological markers show limited application.

2.3.2 BIOCHEMICAL MARKERS

These are a group of proteins formed by the expressed genes that include isozymes, allozymes, and secondary metabolites. A wide array of biochemical markers are extensively used for enzymatic function, investigation of seed protein storage, and isozymes. Allozymes are variant forms of an enzyme whose function is the same but differs structurally. These are contrasting to isozymes, which are enzymes that perform the same function but genes coding the enzymes are placed at different loci. These particulars may be used to estimate population, genetic structure, genetic diversity, gene flow, and population diversity, and comparative study of species out-crossing rates.

2.3.2.1 ADVANTAGES OF BIOCHEMICAL MARKERS

- Major advantages of the above markers are evaluating co-dominance, pleiotropic effects, and absence of epistasis.

- These markers are comparatively inexpensive and a potent method for estimating allele frequencies for specific genes.
- These markers are co-dominant and require a very short fragment of a gene for detection that is why heterozygous to be distinguished from homozygous.
- This method is non-destructive since only a small amount of plant tissue is needed.

2.3.2.2 DISADVANTAGES OF BIOCHEMICAL MARKERS

- The accuracy of this technique is very low because only protein-coding regions were studied and any small change in those regions will change the whole results.
- A limited number of allozymes or isozymes loci are available which's why not useful in genome mapping.
- Only a few isozymes per species approximately 30 corresponding with few markers.
- Relating of distinct samples from contrasting species, loci and laboratories form one of the major concerns being distressed by plant tissue, plant stage, and extraction methodology.

2.3.3 MOLECULAR MARKERS

Molecular markers evoked as Genetic markers comprise a portion of DNA that is related within a specific location in the genome. These markers are being used in different fields of life sciences especially in biotechnology and molecular biology to recognize a specific sequence of DNA in a whole chromosome and genetic variations in plants. These are dominant and co-dominant and have expressed and non-expressed sequences. The study of these markers reveals polymorphisms within DNA sequences between diverse individuals. Different types of mutation like point mutations, deletion, insertions, duplications, and translocations are those differences. There are several molecular markers out of the five markers are used predominately in plant genetics, like RFLP, RAPD, AFLP, Microsatellite, and SNP (Table 2.1).

The advantages of molecular markers are as follows:

- Molecular markers can distinguish polymorphisms that don't result in phenotypic variations.
- These apply to any region of the genome like introns, exons, and regulatory regions.
- The molecular markers exhibit co-dominant nature.
- They don't possess pleiotropic or epistatic effects.

2.3.4 IMPORTANCE OF MOLECULAR TOOLS

The huge loss of a variety of valuable plant species and its adverse impact on the environmental condition and socioeconomic values in the past century showed the importance to conserve plant resources in today's world. Legitimate identification and characterization of plant materials from the environment is a must for the lucrative conservation and sustainable use of plant resources. The main advantages of molecular tools that developed in recent years are easy to assess and less arduous means to appoint known and unknown plant species. Many recent evolutionary and taxonomic problems are solved due to current advancements in molecular tools and techniques that were previously not possible with old phenotypic methods (Table 2.1). Random amplified polymorphic DNA (RAPD), single-nucleotide polymorphisms (SNP), microsatellites DNA bar-coding, and amplified fragment length polymorphism (AFLP) are the molecular techniques currently being used to study plant diversity and plant population genomics. Each of these techniques has its advantages and limitations (Table 2.3). DNA-based molecular markers are made used in the field of embryology, genetics, taxonomy, physiology, etc. These techniques vary in their resolving power to detect genetic variations, the type of data generated, and applicability to specific taxonomic levels (Table 2.3). This book chapter presents a wide picture of the advancement in diverse molecular techniques that can be employed for the analysis of plant population genetics.

2.4 ADVANCEMENT IN MOLECULAR MARKER TECHNIQUES

Due to the discovery of genome-based technologies, molecular marker techniques have become more high-tech. This technique is the amalgamation of several primary techniques with the incorporation of new advanced modifications. There are several new techniques incorporated; here we are discussing some of the key techniques in brief.

2.4.1 SEQUENCE CHARACTERIZED AMPLIFIED REGIONS (SCAR)

SCAR is the technique based on PCR that uses sequence-specific oligonucleotide primers and represents genomic DNA at a definite location in a genome. To use markers distinguished by self-assertive markers (RAPD, AFLP, and so on) for map-based cloning, a single locus should be distinguished unequivocally. Moreover, the discretionary marker strategies are delicate to changes in the response conditions. To overcome any barrier between the capacity to acquire connected markers to quality of interest in a short time and the utilization of these markers for map-based cloning draws near and for routine screening techniques, the SCAR marker method was created and applied. The SCARs are PCR-based markers that speak to genomic DNA pieces at hereditarily characterized loci that are recognized by PCR intensification utilizing grouping explicit oligonucleotide preliminaries (Paran and Michelmore, 1993). Induction of SCARs includes cloning the enhanced results of self-assertive marker procedures and afterward sequencing the two closures of the cloned items. The arrangement is from that point used to plan explicit preliminary sets of 15–30 base pairs which intensify single significant groups of the size like that of the cloned piece. Polymorphism either is held as the presence or nonattendance of intensification of the band or can show up as length polymorphisms convert predominant arbitrary marker loci into codominant SCAR markers. As SCARs are fundamentally characterized hereditarily, they can be utilized both as actual milestones in the genome and as hereditary markers. Codominant SCARs are more informative for genome mapping than dominant arbitrary-primed molecular markers, as they can be utilized to screen pooled genomic libraries by PCR and for actual physical mapping (Chelkowski and Stephen, 2001), characterizing locus explicitness (Paran and Michelmore, 1993) as well as comparative mapping, the study of homologs in between relative plant species.

2.4.2 CLEAVED AMPLIFIED POLYMORPHIC SEQUENCES (CAPS)

The CAPS marker procedure gives an idea to use the DNA sequences of already mapped RFLP markers to create PCR-based markers along these lines removing the monotonous DNA blots (Konori and Nitta, 2005). In this manner, CAPS are moreover known as PCR-RFLP markers (Konieczny and Ausubel, 1993). The CAPS decodes the limitation piece length polymorphisms brought about by single base changes like SNPs, insertion, and deletion, which alter endonucleases responsive destinations in PCR products (Chelkowski and Stephen, 2001; Konieczny and Ausubel, 1993). The CAPS technique is performed, by processing locus-specific PCR products with at least one or more than one restriction enzyme, pursue by the desperation of digested DNA on agarose or polyacrylamide gels. Based on available information in the data bank makes the primer. The CAPS technique is more advance by joining with another single-strand polymorphism that helps find DNA polymorphism. The CAPS markers show codominant and specific for locus so utilized to the reorganization of ploidy (either homo or heterozygous) in plants (Konieczny and Ausubel, 1993). Hence, CAPS are valuable for genotyping; map-based cloning, and molecular chattelizations (Spaniolas et al., 2006; Weiland and Yu, 2003) where identification is impossible based on the sequence. Limitations of this technique are mutations in the genome, after mutation restriction enzyme recognition site is disrupted. To overcome this problem, Michaels and Amasino (1998) develop a new method known as dCAPS (derived cleaved amplified polymorphic sequence). In the dCAPS procedure, SNP is added into the PCR amplicon with the help of a primer containing multiple matches to template DNA (Neff and Chory, 1998). The new PCR product is then exposed to digestion by restriction enzymes and the presence or absence of the SNP is controlled by the subsequent restriction order. This strategy is straightforward, moderately modest, and uses omnipresent technologies like PCR, agarose gel electrophoresis, and restriction enzyme digestion. Another advantage of this technique knows mutations in the plant population and performs position-based cloning of new desired genes in the plant population (Haliassos et al., 1989).

2.4.3 RANDOMLY AMPLIFIED MICROSATELLITE POLYMORPHISM (RAMP)

This technique is the amalgamation of two marker techniques, i.e., microsatellite-based markers, and RAPD markers. Microsatellite markers show a more noteworthy degree of polymorphism having the disadvantage of being more labour demanding. While RAPD markers are low cost and their degree of polymorphism identification is low as compared to microsatellites. To conquer the drawback of these two strategies, RAMP markers were created (Iqbal et al., 2013). This technique includes a radiolabel primer having 5' to 3' repeats used to amplify genomic DNA. The subsequent product is resolved by denaturing PAGE. RAMP markers are practical, reflect higher polymorphism, and have wide dissemination in the genome. They have been effectively applied in different plants for molecular identification (Nadeem et al., 2018).

2.4.4 SEQUENCE-RELATED AMPLIFIED POLYMORPHISM (SRAP)

Li and Quiros (2001) built up a technique that is used for the amplification of open reading frames (ORFs) based on the amplification of two primers. This technique utilizes primer length between 17 and 21 nucleotides of AT or GC rich sequence which amplify internal genomic fragments for polymorphism identification. Primer utilizes the CCGG sequence in the forward AATT in the reverse, and during PCR the annealing temperature in the initial five cycles is set at 35°C. The remaining 35 cycles are run at 50°C. The PCR product is then stacked on gel electrophoresis and DNA bands are pictured through autoradiography. SRAPs are predominant in nature and DNA parts are scored by the presence or nonappearance of a band. This is a basic and proficient marker framework that is broadly utilized in a scope of fields, including map development, genomic, and cDNA fingerprinting (Boopathi, 2013). SRAP is a predominant marker framework that has been effectively applied to research the hereditary varieties in various taxa.

2.4.5 TARGET REGION AMPLIFICATION POLYMORPHISM (TRAP)

The TRAP strategy (Hu and Vick, 2003) is a fast and productive PCR-based method, which uses bioinformatics apparatuses and expressed sequence tag (EST) information base data to create polymorphic markers, around focused targeted genomic sequences. The procedure utilizes two markers (18-nucleotide long) one of the primers is the fixed primer and is originated from the focused-on EST sequence arrangement; the subsequent second primer is an arbitrary primer with one or the other an AT-or GC-rich center to strengthen with an intron or exon. This technique can be utilized to create markers for particular gene sequences, in genotyping and producing markers for marker-assisted breeding (Hu et al., 2008). The method has been successfully utilized in fingerprinting of lettuce cultivars, in assessing hereditary diversification (Alwala et al., 2006), and QTL mapping in a wheat inter-varietal recombinant population.

2.4.6 SINGLE-STRAND CONFORMATION POLYMORPHISM (SSCP)

This is a powerful method for the identification of point mutation (Fukuoka et al., 1994). SSCP is based on mobility shift analysis of single-stranded DNA sequences on neutral polyacrylamide gel electrophoresis to distinguish polymorphisms created by differential folding of single-stranded DNA because of unobtrusive contrasts in DNA succession (Orita et al., 1989). Without a negative strand, the single strand encounters intrastrand base blending, bringing about circles and overlays that give it an extraordinary 3D-structure that can be extensively adjusted because of single base change bringing about differential portability (Orita et al., 1989). The SSCP investigation ends up being an amazing asset for surveying the multifaceted nature of PCR amplicon as the two DNA strands from a similar PCR amplicon (Hayashi, 1992) regularly run independently on SSCP gels, consequently giving two occasions to score a polymorphism and settling inner grouping polymorphisms in some PCR items from indistinguishable spots in the two parental genomes. The PCR-based SSCP investigation is a quick, basic, and touchy procedure for the recognition of different transformations, including single nucleotide replacements, inclusions, and erasures, in PCR-enhanced DNA sections (Hayashi and Yandell, 1993). It is a ground-breaking procedure for quality examination especially for the discovery of point transformations. The major drawback of this technique is costly, labor-intensive, and cannot be automated.

2.5 TRANSPOSABLE ELEMENT-BASED MARKERS

Transposons are versatile hereditary components that have the capacity to alter their position in the genome were identified nearly 60 years back in maize. There are two wide classes of transposable components, each with trademark properties (Finnegan, 1989). For all Class I or retroelements, for example, retrotransposons, short mixed nuclear components, and long mixed nuclear elements, it is the component encoded mRNA, and not simply the component itself that makes transposition intermediates. This implies a new copy in the genome that duplicates the transposon while the first duplicate remaining parts are unblemished at the donor site. In contrast, Class II comprises DNA transposons, which change their area in the genome by a "cut and paste" mechanism (Grzebelus, 2006).

2.5.1 INTER-RETROTRANSPOSON AMPLIFIED POLYMORPHISM (IRAP) AND RETROTRANSPOSON MICROSATELLITE AMPLIFIED POLYMORPHISM (REMAP)

IRAP and REMAP are based on amplification-based marker techniques that have been created dependent on the position of given LTRs inside the genome (Kalendar and Schulman, 2006). These two markers have been grown initially for BARE-1 retrotransposon of the Hordeum variety, which is present in the genome of Barley in various duplicates. The IRAP markers are created by the nearness of two LTRs utilizing outward-confronting preliminaries toughening to LTR target arrangements. In REMAP, amplification between LTRs proximal to basic succession rehashes, for example, constitutive microsatellite produces markers. Both IRAP and REMAP analyze polymorphism in retrotransposon insertion sites in the genome, IRAP among retrotransposons, and REMAP in between retrotransposons and microsatellites (SSRs).

2.5.2 SEQUENCE-SPECIFIC AMPLIFICATION POLYMORPHISM (S-SAP)

The procedure was first used to examine the area of BARE-1 retrotransposons in the grain genome (Manninen et al., 2000). On a basic level, it is a basic alteration of the standard AFLP (amplified fragment length polymorphism). The last amplification is done with retrotransposon-specific and MseI-adaptor specific primers. S-SAP has been broadly used to produce markers to examine the hereditary variety and to get ready linkage maps in a few plants, including the pea, cashew, Medicago, and the wheat (Agarwal et al., 2008).

2.5.3 RETROTRANSPOSON-BASED INSERTION POLYMORPHISM (RBIP)

The procedure was created utilizing the PDR1 retrotransposon in the pea (Flavell et al., 1998). It requires the arrangement data of the 5'and 3' regions flanking the transposon. At the point when a preliminary explicit to the transposon is utilized along with a preliminary intended to strengthen to the flanking portion, they create a product from template DNA having insertion. Then again, primers explicit to both flanking locales amplify a sequence if the addition is missing. Polymorphisms can be distinguished utilizing standard agarose gel electrophoresis or by hybridization with a reference PCR section. Hybridization is more helpful for robotized, high-throughput examination. It is a lot more exorbitant and convoluted than different strategies for distinguishing transposon inclusions.

2.6 ORGANELLE MICROSATELLITES

Plant organelles like mitochondria and chloroplast have been used in the study of population genetics and the phylogeny of plant populations. Chloroplast and mitochondrial genomes show different patterns as compared to nuclear alleles. So far, in addition to nuclear microsatellites, chloroplast and mitochondrial microsatellites have also been studied in population genetic analysis.

2.6.1 MITOCHONDRIAL MICROSATELLITES

Plant mitochondrial DNA (mtDNA) is extremely unique, with the biggest and the least quality thickness among eukaryotes. Its size ranges somewhere in the range of 200 and 2500 kb and comprises various rehashed components and introns (Kozik et al., 2019). Plant mtDNA is bigger when contrasted with the animal mtDNA and is portrayed by atomic heterogeneity seen as gatherings of round chromosomes that vary in size and plenitude. The development pace of mtDNA markers is moderate, which is deplorable for plant populations scholars; these markers have exceptionally restricted application in population genetics. Mitochondrial DNA markers display numerous impediments like the way that they speak to just a solitary locus; vulnerability in the genealogical investigation can be expanded because of the expanded likelihood of missing connections in mitochondrial haplotypes and underestimation of hereditary variation (Dowell, 2014).

2.6.2 CHLOROPLAST MICROSATELLITES (CPSSRS)

It is hard to distinguish enough sequence variations because of the lesser change rates that describe the chloroplast genome. As opposed to this, cpSSRs furnish higher polymorphism levels with effectively genotyping, which has made them an exceptionally small bunch and mainstream marker for population hereditary investigations (Provan et al., 2001). cpSSRs ordinarily contain a single nucleotide motif which is repeated by 8–15 times. The polymorphism level in cpSSRs is very evolving across species and loci. Two significant highlights recognizing the cpSSRs from atomic microsatellites are (1) chloroplasts are acquired uniparentally and (2) the chloroplast chromosome is a non-recombinant atom due to which all cpSSRs loci are connected. CpSSRs have been effectively applied in agribusiness and essential plant sciences (Ebert and Peakall, 2009).

2.7 SEQUENCING MARKERS

There are four types of sequencing markers are present that are given below

(a) Sanger method of sequencing

(b) Pyro-sequencing
(c) Next-generation sequencing (NGS)
(d) Genotyping by sequencing (GBS)

2.7.1 SANGER METHOD OF SEQUENCING

This method was invented by Fred Sanger and Coulson in 1977 in which they used modified bases known as dideoxynucleotides. The basic principle of this method is single-stranded DNA molecules having length differentiation of a single nucleotide that can be separated from each other with the help of PAGE. It involves five important components like a primer, DNA template to be sequenced, a DNA polymerase enzyme, ddNTPs that are used as terminators each labelled with a different color of the dye, dNTPs (A, G, T, and C). The termination of DNA synthesis reaction occurs when ddNTPs are added to the growing oligonucleotide chain that resulted in stemming products of varying lengths with a suitable ddNTP at their 3'-terminus. The final product is extracted based on size using PAGE and the terminal ddNTPs are utilized to the DNA sequence of the DNA template strand. This method of sequencing includes four steps are denaturation, primer attachment and extension of bases, termination and final step is polyacrylamide gel electrophoresis.

Advantages of Sanger method of sequencing are

- High variation measurement
- Reproducibility is high
- Low quality of DNA required

Disadvantages of Sanger method of sequencing are

- Time-consuming
- Procedure is costly
- Genome coverage is low
- Detection of low-level polymorphism below the species level

2.7.2 PYROSEQUENCING

Pyrosequencing is a non-electric and bioluminescence type of sequencing technique that estimates the secretion of inorganic pyrophosphate that further converts into visible light by utilizing a series of enzymatic reactions. The principle of Pyrosequencing was invented by Bertil Pettersson, Pal Nyren, and Mathias Uhlen in 1996 at the Royal Institute of Technology in Stockholm by using the solid phase sequencing method. The enzyme luciferase produced light that further catalyzed reaction is captured by a charge-coupled device (CCD) camera and shown as a peak in a computer program. In this technique, the addition of deoxynucleotide to the end of the growing strand is detectable because it is accompanied by the release of a flash of light. The total number of nucleotides incorporated is propositional to each signal produced (Fig. 2.2).

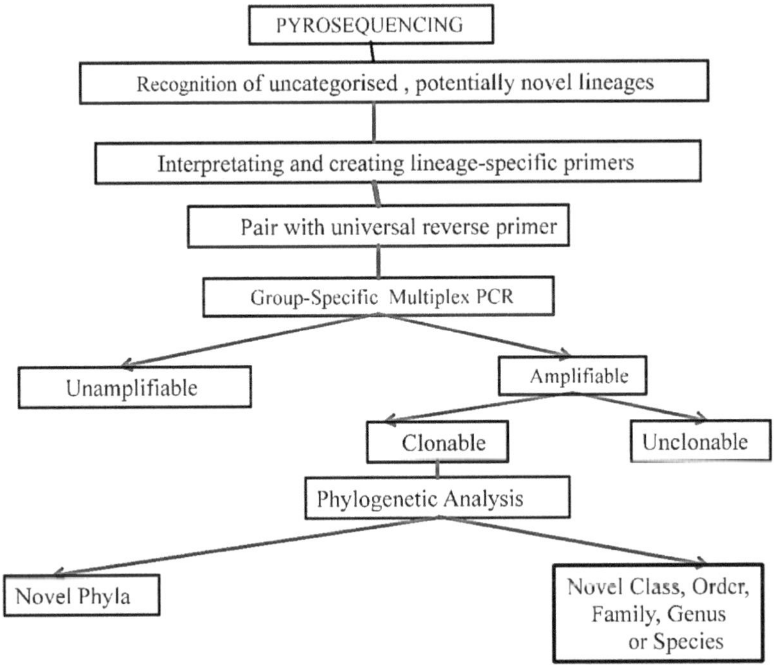

FIGURE 2.2 Flowchart of the combined Pyrosequencing-Sanger approach for novel lineage identification.

2.8 NEXT-GENERATION SEQUENCING

Next-generation sequencing was invented due to the demand for low-cost sequencing methods. This is the main reason for the discovery of next-generation sequencing. This is the method through which hundreds to million times to billion times of DNA are produced per run.

Advantages of next-generation sequencing are

- Cost-effective
- More accurate procedure among older sequencing techniques like Pyrosequencing and Sanger method.
- Useful in whole-genome sequencing for detection of a maximum number of SNPs.
- Useful for sequencing of old DNA samples.
- Useful in the field of metagenomics.

2.8.1 GENOTYPING BY SEQUENCING

Genotyping by sequencing is also adverted as GBS in the field of genetic sequencing. This method was developed to ferret out SNPs for carrying out genotyping studies like genome-wide association studies (GWAS). GBS is a simple, wholesome, and affordable technique for SNP mapping and its easy detection. It was mainly invented for different types of species having a complicated and complex type of genome. GBS uses the procedure of digestion with the help of restriction endonuclease to decrease the complexity of genome and genotype DNA samples. First, this method was discovered for the investigation of high-resolution association maize crops. This sequencing method was discovered in Buckler's lab with the help of the Illumina NGS technique.

There are two types of GBS based on ion PGM system usage are:

- Cutting through restriction enzyme: This sequencing method is used for marker-assisted selection (MAS) programs for selecting new markers.
- Multiplex enrichment PCR: In this sequencing technique, specific PCR primers are chosen for the amplification of DNA strands. A complete set of SNPs are identified in this method instead of using restriction enzymes for the digestion of DNA strands in the Sanger sequencing method.

The advantages of next-generation sequencing are

- Lowered the cost of sequencing

- Useful for a wide range of diversity including mostly large genome species
- Useful for a depiction of population studies, germplasm, and plant breeding of different types of crops
- Curtail a great extent of SNPs used in genotyping and genetic studies
- Decreases handling of samples
- Decreases PCR process and purification technique set

2.8.2 RNA-BASED MARKER

Genetic expression control can be synchronized by different factors like developmental programming and biological processes. The study of differential patterns of gene expression can be gathered for differential RNA study. PCR-based marker techniques like RAP-PCR and cDNA-AFLP are used for the study of distinctive RNA study by finicky amplification of cDNAs. There are three types of RNA-based markers viz., (1) CDNA-SSCP (2) RAP-PCR (3) CDNA-AFLP.

2.8.2.1 CDNA-SSCP

This technique is used for the silencing process of global frequency in natural and synthetic polyploids. In this method, PCR products of SSCP analysis are used for the evaluation of highly similar homologous gene pairs of polyploids. The separation of duplicated transcripts comprising 99% sequence similarity is mostly known to be the most authentic and reliable technique according to the replicated tests (Cronn and Adams, 2003). This sequencing method is used to acquire silencing into the global frequency in both natural and synthetic polyploids.

2.8.2.2 RNA FINGERPRINTING BY ARBITRARILY PRIMED PCR (RAP-PCR)

This technique is used for the silencing process of RNA samples using arbitrarily selected primer followed by PCR amplification of the cDNA population. The requirement of RNA in this method is in nanograms of the total RNA sample. RNA detection from the same tissue from different individuals and RNAs from different tissue from the same individual by differential PCR fingerprints were obtained. Due to sequence polymorphisms, the individual-specific differences were disclosed. It is useful for the revelation of tissue-specific differences and differential gene expression study. It is also applicable for the genetic mapping of genes.

2.8.2.3 CDNA -AFLP

The cDNA-AFLP is an innovative RNA fingerprinting technique to show differential expressed genes. The first step of this methodology is the cutting of DNAs by two restriction enzymes. The second step is to ligate to two oligonucleotides adapters followed by PCR amplification using additional selective nucleotides (Bachem et al., 1996). This technique is more reproducible and more stringent than previous techniques like RAP-PCR. The requirement of sequence information is not compulsory in CDNA-AFLP. It is an important instrument for the recognition of innovative genes that is process-related. A major application of this technique is to identify stress-regulated genes.

2.9 CONCLUSION

This paper briefly discusses developments and advancements of molecular tools in the area of plant population genetics. The regular evolution in molecular marker technologies has evolved in the development of ultra-high-throughput genotyping platforms. Even so, low-throughput molecular markers such as SSRs are quite essential for tagging specific genomic regions in molecular breeding programs. For genome-wide marker screening, SNP markers are most preferred for the development of high-throughput genotypic platforms. It appears that the recently discovered NGS-based molecular marker system may replace the array-based high-throughput marker systems in the forthcoming future, especially when costs are reducing and throughput is enhancing for the NGS technologies. These marker systems in the future may be very useful for enhancing the extension of modem genetics and breeding approaches such as GWAS and GWS that are still used as a beginning procedure in plant systems for crop improvement. Recent molecular technologies also have an essential impact in many fundamental and applied fields of parasitology. Molecular diagnostic tools as nucleotide sequencing have been broadly utilized through continuous advancement in molecular biology and biotechnology helpful in the diagnosis of genetic and microbial diseases in plants. The development of molecular diagnostic assays with the help of molecular tools in the field of biotechnology is an important discovery to a critical disease response in plants. In the forthcoming days, advancement in the molecular tool in the area of plant population genetics may reduce obstacles like the choice of technology and equipment, and other issues like cost-effectiveness, reproducibility, and accuracy. Thus, the coming era of molecular biotechnology will be innovatory, which would change the way of diagnostics and treatment of plant diseases and the process of analysis of plant population genetics.

TABLE 2.1 Differentiation of Different Characteristics of Most Frequently Used Molecular Markers Techniques.

Molecular markers	RFLP	RAPD	AFLP	SSR	CAPS	SCAR	IRAP	RAMP	SSCP	SNP
Degree of Polymorphism	M	M	M	M	L	M	M	M	L	H
Locus specificity	C	D	D	C	C	C	D	D	C	C
Dominance(D)/ Co-dominance (C)	C	D	D	C	C	C	D	D	C	C
Ease of replication	H	L	H	M	H	H	H	M	M	H
Abundance	H	H	H	M	L	L	H	M	M	H
Automation	N	Y	Y	Y	Y	Y	Y	Y	N	Y
Cost per assay	H	L	M	L/M	M	L	L	M	H	L
Technical requirement	H	L	M	L/M	H	M	H	H	H	M

H, high; M, medium; L, Low; Y, Yes; N, No.

TABLE 2.2 Acronyms Commonly Used for a Different Type of Molecular Markers.

AFLP	Amplified fragment length polymorphism
AP-PCR	Arbitrarily primed PCR
ARMS	Amplification Refractory Mutation System
ASAP	Arbitrary Signatures from Amplification
ASH	Allele-Specific Hybridization
ASLP	Amplified Sequence Length Polymorphism
ASO	Allele Specific Oligonucleotide
CAPS	Cleaved Amplification Polymorphic Sequence
CAS	Coupled Amplification and Sequencing
DAF	DNA Amplification Fingerprint
DGGE	Denaturing Gradient Gel Electrophoresis
GBA	Genetic Bit Analysis
IRAO	Inter-Retrotrasposon Amplified Polymorphism
ISSR	Inter-Simple Sequence Repeats
ISTR	Inverse Sequence-Tagged Repeats
MP-PCR	Microsatellite-Primed PCR
OLA	Oligonucleotide Ligation Assay
RAHM	Randomly Amplified Hybridizing Microsatellites
RAMPs	Randomly Amplified Microsatellite Polymorphisms
RAPD	Randomly Amplified Polymorphic DNA
RFLP	Restriction Fragment Length Polymorphism
REF	Restriction Endonuclease Fingerprinting
REMAP	Retrotransposon-Microsatellite Amplified Polymorphism
SNP	Single Nucleotide Polymorphism
SSLP	Single Sequence Length Polymorphism
STS	Sequence-Tagged-Site
STMS	Sequence-Tagged Microsatellite Site
SSR	Simple Sequence Repeats
SSCP	Single Strand Conformation Polymorphism
SPAR	Single Primer Amplification Reaction
SPLAT	Single Polymorphic Amplification Test
S-SAP	Sequence-Specific Amplification Polymorphisms
TGGE	Thermal Gradient Gel Electrophoresis
VNTR	Variable Number Tandem Repeats
RAMS	Randomly Amplified Microsatellites
TD	Transposable Display
RBIP	Retrotransposan -Based Amplified Polymorphism

TABLE 2.3 Comparison of Different Aspects of Frequently Used Molecular Marker Techniques.

Molecular markers	Abundance	Reproducibility	Degree of polymorphism	Locus specificity	Technical requirement	Quantity of DNA requirement
RFLP	High	Medium	Medium	Yes	High	High
RAPD	High	Low	Medium	No	Low	Low
SSCP	Low	Medium	Low	Yes	Medium	Low
SSR	Medium	Medium	Medium	No	Medium	Low
CAPS	Low	High	Low	Yes	High	Low
SCAR	Low	High	Medium	Yes	Medium	Low
AFLP	High	High	Medium	No	Medium	Medium
RAMPO	Medium	Medium	Medium	Yes	High	Low
REMAP	High	High	Medium	Yes	High	Low
SNP	High	High	High	High	Variable	High
RETROTRANSPOSAN	High	High	Very high	High	Low	High

RFLP, restriction fragment length polymorphism; RAPD, random amplified polymorphic DNA; SSCP, single-strand conformational polymorphism; SSR, simple sequence repeats; CAPS, cleaved amplified polymorphic sequence; SCAR, sequence characterized amplified region; AFLP, amplified fragment length polymorphism; RAMP, randomly amplified microsatellite polymorphisms; REMAP, retrotransposon microsatellite amplified; SNP, single nucleotide polymorphism.

KEYWORDS

- **molecular tools**
- **plant population genetics**
- **assessment of diversity**

REFERENCES

Agarwal, M.; Shrivastava, N.; Padh, H. Advances in Molecular Marker Techniques and Their Applications in Plant Sciences. *Plant Cell Rep.* **2008,** *27* (4), 617–631.

Akihiro, T.; Umezawa, T.; Ueki, C.; Lobna, B. M.; Mizuno, K.; Ohta, M.; Fujimura, T. Genome-wide cDNA-AFLP Analysis of Genes Rapidly Induced by Combined Sucrose and ABA Treatment in Rice Cultured Cells. *FEBS Lett.* **2006,** *580* (25), 5947–5952.

Alwala, S.; Suman, A.; Arro, J. A.; Veremis, J. C.; Kimbeng, C. A. Target Region Amplification Polymorphism (TRAP) for Assessing Genetic Diversity in Sugarcane Germplasm Collections. *Crop Sci.* **2006,** *46* (1), 448–455.

Bachem, C. W.; Van Der Hoeven, R. S.; De Bruijn, S. M.; Vreugdenhil, D.; Zabeau, M.; Visser, R. G. Visualization of Differential Gene Expression Using a Novel Method of RNA Fingerprinting Based on AFLP: Analysis of Gene Expression during Potato Tuber Development. *Plant J.* **1996,** *9* (5), 745–753.

Bachem, C. W.; Oomen, R. J.; Visser, R. G. Transcript Imaging with cDNA-AFLP: A Step-by-Step Protocol. *Plant Mol. Biol. Rep.* **1998,** *16* (2), 157–157.

Bardakci, F. Random Amplified Polymorphic DNA (RAPD) Markers. *Turk. J. Biol.* **2001,** *25* (2), 185–196.

Boopathi, N. M. *Genetic Mapping and Marker Assisted Selection. Basics, Practice and Benefits*; Springer: New York, 2013; 293 p. http://dx.doi.org/10.1007/978-81-322-0958-4

Boopathi, N. M. *Genetic Mapping and Marker Assisted Selection*; Springer: New York, **2020**; 504 p.

Boopathi, N. M.; Sathish, S.; Kavitha, P.; Dachinamoorthy, P.; Ravikesavan, R. Molecular Breeding for Genetic Improvement of Cotton (Gossypium spp.). In *Advances in Plant Breeding Strategies: Breeding, Biotechnology and Molecular Tools*; Jameel, A-K., Jain, M., Johnson, M., Dennis, V., Eds.; Springer: Cham, 2015; pp 613–645.

Botstein, D.; White, R. L.; Skolnick, M.; Davis, R. W. Construction of a Genetic Linkage Map in Man Using Restriction Fragment Length Polymorphisms. *Am. J. Hum. Genet.* **1980,** *32* (3), 314.

Burgess, T.; Wingfield, M. J.; Wingfield, B. W. Simple Sequence Repeat Markers Distinguish among Morphotypes of *Sphaeropsissapinea. Appl. Environ. Microbiol.* **2001,** *67* (1), 354–362.

Cronn, R. C.; Adams, K. L. Quantitative Analysis of Transcript Accumulation from Genes Duplicated by Polyploidy Using cDNASSCP. *Biotechniques* **2003,** *34* (4), 726–734.

Davey, J. W.; Hohenlohe, P. A.; Etter, P. D.; Boone, J. Q.; Catchen, J. M.; Blaxter, M. L. Genome-Wide Genetic Marker Discovery and Genotyping Using Next-Generation Sequencing. *Nat. Rev. Genet.* **2001,** *12* (7), 499–510.

Dowell, D. R. *NextGen Genealogy: The DNA Connection*; ABC-CLIO, 2014.

Hill, R. *Finding Family: My Search for Roots and the Secrets in My DNA*; Workman Publishing, 2017.

Ebert, D.; Peakall, R. O. D. Chloroplast Simple Sequence Repeats (cpSSRs): Technical Resources and Recommendations for Expanding cpSSR Discovery and Applications to a Wide Array of Plant Species. *Mol. Ecol. Resour.* **2009,** *9* (3), 673–690.

Elshire, R. J.; Glaubitz, J. C.; Sun, Q.; Poland, J. A.; Kawamoto, K.; Buckler, E. S.; Mitchell, S. E. A Robust, Simple Genotyping-by-Sequencing (GBS) Approach for High Diversity Species. *PloS One 6* (5), e19379.

Finnegan, D. J. Eukaryotic Transposable Elements and Genome Evolution. *Trends Genet.* **1989,** *5*, 103–107.

Fukuoka, H.; Kawatani, M.; Hisamitsu, T.; Takeshige, C. Cutaneous Hyperalgesia Induced by Peripheral Injection of Interleukin-1β in the Rat. *Brain Res.* **1994,** *657* (1–2), 133–140.

Grzebelus, D. Transposon Insertion Polymorphism as a New Source of Molecular Markers. *J. Fruit Ornam. Plant Res.* **2006,** *14*, 21.

González, A.; Wong, A.; Delgado Salinas, A.; Papa, R.; Gepts, P. Assessment of Inter Simple Sequence Repeat Markers to Differentiate Sympatric Wild and Domesticated Populations of Common Bean. *Crop Sci. 45* (2), 606–615.

Guarino, L. Approaches to Measuring Genetic Erosion. In *Technical Meeting on the Methodology of the FAO World Information and Early Warning System on Plant Genetic Resources*; Prague (Czech Republic), 1999; pp 21–23.

Haliassos, A.; Chomel, J. C.; Grandjouan, S.; Kruh, J.; Kaplan, J. C.; Kitzis, A. Detection of Minority Point Mutations by Modified PCR Technique: A New Approach for a Sensitive Diagnosis of Tumor-Progression Markers. *Nucl. Acids Res.* **1989,** *17* (20), 8093–8099.

Hamrick, J. L.; Godt, M. J. W. Allozyme Diversity in Cultivated Crops. *Crop Sci.* **1997,** *37* (1), 26–30.

Hayashi, K. PCR-SSCP: A Method for Detection of Mutations. *Genet. Anal. Biomol. Eng.* **1992,** *9* (3), 73–79.

Hayashi, K.; Yandell, D. W. How Sensitive Is PCR-SSCP. *Human Mutation* **1993,** *2* (5), 338–346.

He, J.; Zhao, X.; Laroche, A.; Lu, Z. X.; Liu, H.; Li, Z. Genotyping-by-Sequencing (GBS), an Ultimate Marker-Assisted Selection (MAS) Tool to Accelerate Plant Breeding. *Front. Plant Sci.* **2014,** *5*, 484.

Hu, J.; Vick, B. A. Target region Amplification Polymorphism: A Novel Marker Technique for Plant Genotyping. *Plant Mol. Biol. Rep.* **2003,** *21* (3), 289–294.

Iqbal, A. M.; Qadri, H.; Lone, A. A.; Wani, S. A.; Ali, G.; Bhat, A.; Dar, Z. A. Marker-Assisted Selection in Crop Plants. *LS: Int. J. Life Sci.* **2013,** *2* (1), 44–57.

Joshi, S. P.; Ranjekar, P. K.; Gupta, V. S. Molecular Markers in Plant Genome Analysis. *Curr. Sci.* **1999,** 230–240.

Karp, A.; Seberg, O. L. E.; Buiatti, M. Molecular Techniques in the Assessment of Botanical Diversity. *Ann. Bot.* **1996,** *78* (2), 143–149.

Karp, A. Molecular Tools in Plant Genetic Resources Conservation: A Guide to the Technologies (No. 2). *Bioversity Int.* **1997,** 215.

Kalendar, R.; Schulman, A. H. IRAP and REMAP for Retrotransposon-Based Genotyping and Fingerprinting. *Nat. Protocols* **2006,** *1* (5), 2478–2484.

Konieczny, A.; Ausubel, F. M. A Procedure for Mapping Arabidopsis Mutations Using Co-dominant Ecotype Specific PCR Based Markers. *Plant J.* **1993,** *4* (2), 403–410.

Kozik, A.; Rowan, B. A.; Lavelle, D.; Berke, L.; Schranz, M. E.; Michelmore, R. W.; Christensen, A. C. The Alternative Reality of Plant Mitochondrial DNA: One Ring Does Not Rule Them All. *PLoS Genetics* **2019,** *15* (8), 1008373.

Lew, K. *Genetic Ancestry Testing*; Enslow Publishing, LLC, 2018.

Li, G.; Quiros, C. F. Sequence-Related Amplified Polymorphism (SRAP), a New Marker System Based on a Simple PCR Reaction: Its Application to Mapping and Gene Tagging in *Brassica. Theor. Appl. Genet.* **2001,** *103* (2–3), 455–461.

Liu, H.; Bayer, M.; Druka, A.; Russell, J. R.; Hackett, C. A.; Poland, J.; Waugh, R. An Evaluation of Genotyping by Sequencing (GBS) to Map the Breviaristatum-e (ari-e) Locus in Cultivated Barley. *BMC Genomics* **2014,** *15* (1), 1–11.

Liu, L. W.; Zhao, L. P.; Gong, Y. Q.; Wang, M. X.; Chen, L. M.; Yang, J. L.; Wang, L. Z. DNA Fingerprinting and Genetic Diversity Analysis of Late-Bolting Radish Cultivars with RAPD, ISSR and SRAP Markers. *Sci. Horticult.* **2008,** *116* (3), 240–247.

Liang, P.; Pardee, A. B. Differential Display of Eukaryotic Messenger RNA by Means of the Polymerase Chain Reaction. *Science* **1992,** *257* (5072), 967–971.

Mao, C.; Yi, K.; Yang, L.; Zheng, B.; Wu Y.; Liu F.; Wu P. Identification of Aluminium-Regulated Genes by cDNA-AFLP in Rice (Oryza sativa L.): Aluminium-Regulated Genes for the Metabolism of Cell Wall Components. *J. Exp. Bot.* **2004,** *55* (394), 137–143.

McDermott, J. M.; Brandle, U.; Dutly, F.; Haemmerli, U. A.; Keller, S.; Muller, K. E.; Wolfe, M. S. Genetic Variation in Powdery Mildew of Barley: Development of RAPD, SCAR, and VNTR Markers. *Phytopathology* **2004,** *84* (11), 1316–1321.

Michaels, S. D.; Amasino, R. M. A Robust Method for Detecting Single Nucleotide Changes as Polymorphic Markers by PCR. *Plant J.* **1998,** *14* (3), 381–385.

Miyata, M.; Yamamoto, T.; Komori, T.; Nitta, N. Marker-Assisted Selection and Evaluation of the QTL for Stigma Exsertion Under Japonica Rice Genetic Background. *Theoret. Appl. Genet.* **2007,** *114* (3), 539–548.

Mullis, K.; Faloona, F.; Scharf, S.; Saiki, R. K.; Horn, G. T.; Erlich, H. Specific Enzymatic Amplification of DNA In Vitro: The Polymerase Chain Reaction. In *Cold Spring Harbor Symposia on Quantitative Biology*; Cold Spring Harbor Laboratory Press, 1986; pp 263–273.

Nadeem, M. A.; Habyarimana, E.; Çiftçi, V.; Nawaz, M. A.; Karaköy, T.; Comertpay, G.; Ercişli, S. Characterization of Genetic Diversity in Turkish Common Bean Gene Pool Using Phenotypic and Whole-genome DArTseq-generated silicoDArT Marker Information. *PloS One* **2018,** *13* (10), 0205363.

Neff, M. M.; Chory, J. Genetic Interactions between Phytochrome A, Phytochrome B, and Cryptochrome 1 during Arabidopsis Development. *Plant Physiol.* **1998,** *118* (1), 27–35.

Orita, M.; Suzuki, Y.; Sekiya, T.; Hayashi, K. Rapid and Sensitive Detection of Point Mutations and DNA Polymorphisms Using the Polymerase Chain Reaction. *Genomics* **1989,** *5* (4), 874–879.

Paran, I.; Michelmore, R. W. Development of Reliable PCR-Based Markers Linked to Downy Mildew Resistance Genes in Lettuce. *Theor. Appl. Genet.* **1993,** *85* (8), 985–993.

Reed, D. H.; Frankham, R. Correlation between Fitness and Genetic Diversity. *Conserv. Biol.* **2003,** *17* (1), 230–237.

Szövényi, G.; Primack, RB. Essentials of Conservation Biology. *Commun. Ecol.* **2011,** *12* (1), 142–142.

Spooner, D. Molecular Markers for Genebank Management (No. 10). Bioversity International, 2005.

Southern, E. M. Detection of Specific Sequences among DNA Fragments Separated by Gel Electrophoresis. *J. Mol. Biol.* **1975,** *98* (3), 503–517.

Spaniolas, S.; May, S. T.; Bennett, M. J.; Tucker, G. A. Authentication of Coffee by Means of PCR-RFLP Analysis and Lab-on-a-chip Capillary Electrophoresis. *J. Agricultural Food Chem.* **2006,** *54* (20), 7466–7470.

Volis, S.; Mendlinger, S.; Turuspekov, Y.; Esnazarov, U.; Abugalieva, S.; Orlovsky, N. Allozyme Variation in Turkmenian Populations of Wild Barley, *Hordeum spontaneum* Koch. *Ann. Bot.* **2001,** *87* (4), 435–446.

Weiland, J. J.; Yu, M. H. A Cleaved Amplified Polymorphic Sequence (CAPS) Marker Associated with Root Knot Nematode Resistance in Sugarbeet. *Crop Sci.* **2003,** *43* (5), 1814–1818.

Welsh, J.; Chada, K.; Dalal, S. S.; Cheng, R.; Relph, D.; McClelland, M. Arbitrarily Primed PCR Fingerprinting of RNA. *Nucl. Acids Res.* **1992,** *20* (19), 4965–4970.

Welsh, J.; McClelland, M. Fingerprinting Genomes Using PCR with Arbitrary Primers. *Nucl. Acids Res.* **1990,** *18* (24), 7213–7218.

Williams, J. G.; Kubelik, A. R. Livak, K. J.; Rafalski, J. A.; Tingey, S. V. DNA Polymorphisms Amplified by Arbitrary Primers Are Useful as Genetic Markers. *Nucl. Acids Res.* **1990,** *18* (22), 6531–6535.

CHAPTER 3

Application of Next-Generation Sequencing in Plant Molecular Ecology

HARSHIT KUMAR SONI[1] and APARNA SINGH[2,*]

[1]*Department of Zoology, Government Science College, Pandhurna 480334, Madhya Pradesh, India*

[2]*Department of Plant Sciences, Avvaiyar Government College for Women, Karaikal 609602, Puducherry, India*

[*]*Corresponding author. E-mail: aparnasingh@dhtepdy.edu.in, aparna295@gmail.com*

ABSTRACT

How organisms respond to biotic and abiotic changes in the natural environment has always been enigmatic. Short-term responses are considered ecological responses and if they are selected for the long term, they have evolutionary importance. In this context, the holobiont is thought to interact with the environmental conditions, and plant with the associated genomes acts as a consortium which responds to environmental changes and evolves as a whole. Advanced techniques like next generation sequencing (NGS) are useful in determining community and ecosystem responses through organismic level response in the form of variations in genes and expression of their alleles and/or epialleles. This chapter outlines how new trends in the field of sequencing can shape the future of omics studies of nonmodel organisms in an ecological context.

Plant Ecogenomics. Peerzada Arshid Shabir, PhD, Peerzada Yasir Yousuf, PhD & Khalid Rehman Hakeem, PhD (Eds.)

3.1 INTRODUCTION

For a long, interactions of the organism with the environment and other organisms have been considered of great importance in shaping the future of a community. It also provides the platform for natural selection in a path to the successful evolution of life forms. Interactions between organisms are conceivable in the form of different phenomena, such as competition, pathogenesis, symbiosis, and kinship. Also, environmental interactions integrate the effect of abiotic factors on life. Interactions are found to be mostly driven by the phenotypes. However, the phenotype is just an expression of the DNA sequences in coordination with different modifications of/by the proteins. The phenotype of whole life form in isolation largely depends upon a nucleic acid molecule and its sequence but the evolution of the same is a result of hierarchical interaction of phenotypes resulting from different DNA sequences contained in different organisms and how the action of genes responds to ontogenic and environmental variations. This chapter revisits the ecological interactions in terms of genes and molecular mechanisms driven by these specialized sequences which are chemical in origin but regulated by a natural phenomenon in response to the environment.

Advanced trends in genomics have devised the detailed molecular mechanisms of biological processes, however, gene function can be precisely determined only in close relation to the environment. The expression of a gene largely depends upon its regulation which simultaneously depends on the cues from interactions with environmental factors. Ironically, when organisms are studied in vitro, they are grown in controlled conditions, which are isolated from their environmental niche. This facilitates the study of isolated functions and mechanisms, masking the effect of environmental factors. This is evident from an earlier study about the flowering time, one of the widely researched genetic mechanisms, in which mapped Quantitative Trait Loci (QTLs) for bolting in controlled environments did not show similar associations in a natural environment. Out of the QTLs mapped in recombinant inbred lines of *Arabidopsis thaliana,* some QTLs were not detected in the natural environment and others had been redetected in a specific set of environments. Additionally, some novel QTLs were detected in natural conditions (Weinig et al., 2002). Such findings of a best-studied trait illustrate the complexity of environmental dependence. Another question that arises while studying functional genomics is that how these interactions with biotic and abiotic factors affect a particular genetic mechanism in an organism? Furthermore, in a separate study where three different transgenic lines expressing three different transgenes in *Nicotiana attenuata* to improve herbivory response against *Manduca sexta* were planted in the natural environment to assess the response of herbivore community, all the three lines showed similar results as shown in vitro concerning herbivore damage. Although they were attacked by different herbivores. This shows the shift of natural selection in response to gene expression change (Kessler et al., 2004). The additional complexity revealed in these cases brings more importance for whole-genome research in an ecological context. Next generation sequencing in nonmodel species will provide the platform to support this conceptual understanding.

3.2 IMPLICATION OF THE WHOLE-GENOME APPROACHES IN ECOLOGICAL STUDIES

The whole genome of an organism can be represented as an organization of haplotypes because of the association between alleles occurring together more often than by chance (Wall and Pritchard, 2003). Within these regions, as a consequence of linkage disequilibrium, the recombination frequency is very low (Goldstein and Weale, 2001). Estimation of the length of these linkage disequilibrium blocks is in the range of 1–100 kbp in natural populations (McVean, 2007). Therefore, on average, a particular organism possesses around 10^5–10^7 linkage disequilibrium blocks. The study of different haplotypes in an organism requires thousands of molecular markers to be developed. Next-generation sequencing technology provides a simple means to assess the sequence diversity in an organism.

A very initial study dealing with the whole-genome approach, cDNA microarray of mycorrhizal and nonmycorrhizal roots of *Medicago truncular* showed upregulation of genes related to stress and defense responses at first, followed by their downregulation (Massoumou et al., 2007). The genes involved in signaling pathways for nodulation have also been found upregulated and their upregulation showed temporal correlation with the root colonization which shows that response toward symbiosis is conserved (Mohammadi-Dehcheshmeh et al., 2018). Interestingly, plant–fungal combinations are specific for a particular environmental condition and are affected by nutrient cycles, and in turn, variations in nutrient-uptake kinetics play an important role in variations of gene expressions resulting in functional diversity (Burleigh et al., 2002). These studies clearly show that plants first respond in a defensive way to the fungal attack but subsequently allow the fungus to colonize the roots. Additionally, plants can differentiate their responses toward biotic and abiotic factors during symbiotic interaction.

In the present context with the evolution of the sequencing approach and reduction in the cost of sequencing, the future of studying the genome, transcriptome, and epigenome of any nonmodel species looks promising (Metzker, 2010). Within the last few years, the cost of sequencing has been drastically reduced with the advancement of technology. Many model organisms have been successfully sequenced and genome assemblies have been properly annotated. Transcriptomes and genomes sequenced from nonmodel species can be aligned on a properly annotated reference genome of the model organism. Although to eliminate the necessity of a reference assembly, many methods have been developed to perform *de novo* analyses (Forouzan et al., 2017).

3.3 NEXT-GENERATION SEQUENCING TECHNIQUE

NGS is considered a revolutionary technique for generating whole-genome data. Until the 1970s, the problem with purification and limited technological developments had hindered the whole genomic studies as the sequencing of the genome was not possible. Also, the generation of big data was stalled with limited data storage capacity. With the development of chemical and biological technologies along with the development of electronics, computation, and big data analysis, the sequencing of bigger nucleic acid flourished (Attwood, 2011). By the start of the 1970s, the sequencing only up to 100 nt was possible. By the end of the 1970s, the development of a method for sequencing by synthesis (Sanger, Nicklen et al., 1977) as well as technologies developed for sequencing by cleavage (Maxam and Gilbert, 1977) allowed the adequate and precise sequencing of longer sequences. For the first time, 5386 bases long single-stranded bacteriophage φ X174 was sequenced completely using Sanger sequencing (Sanger, Air et al., 1977; Sanger et al., 1978). These methods for sequencing were called first-generation sequencing.

DNA sequencing technologies invented by Maxam and Gilbert as well as Sanger were a breakthrough and showed that the sequencing of long fragments as possible. With this, scientists around the world started thinking that decoding the DNA sequence of an organism was important to gain an insight into its genetic makeup and to understand the function of genes and their regulation. Furthermore, it was helpful to decipher various questions related to the origin of life, its organization, and genetic diseases (Gupta and Gupta, 2014). Although available sequencing technology was not high throughput and economical (Kumar et al., 2019). Hence, it was required to develop new sequencing technologies, which would be high throughput as well as economical. These methods were commonly referred to as next-generation sequencing methods, developed and commercialized by various biotech companies, such as Thermo Fisher Scientific, Applied Biosystems, Roche Life Sciences, and Illumina. These methods have transformed the landscape of nucleic acid sequencing. Various platforms used for sequencing and their manufacturers have been listed in Table 3.1.

TABLE 3.1 List of Various Sequencing Platforms and Their Principles.

S. No.	Platform	Basic principle	Sequencing method
1	Roche/454 sequencing	Sequencing by synthesis	Pyrosequencing/detection of pyrophosphate
2	Ion Torrent	Sequencing by synthesis	Proton detection
3	Illumina (Solexa) sequencing	Sequencing by synthesis	Detection of fluorescence and use of reversible terminator
4	SOLiD sequencing	Sequencing by ligation	High fidelity ligation with two base encoding

Broadly, the methods for next-generation sequencing which are used widely these days are part of the second-generation sequencing. The sequence data produced by NGS is made up of billions of short DNA sequences, which were called reads. These reads are in the range of 50–300 bp long. Short reads sequencing methods were broadly divided into two categories—sequencing by ligation (SBL) and sequencing by synthesis (SBS) (Kumar et al., 2019). SBS utilizes the DNA polymerization capacity of DNA polymerase. DNA strands are synthesized simultaneously using nucleotide or short oligos, which are either modified with tag or introduced one by one so that the incorporation of bases can be recognized by the appropriate detection technology. SBL methods use the capacity of DNA ligase to ligate DNA sequence and its mismatch sensitivity.

3.3.1 ROCHE/454 SEQUENCING

This NGS platform was first developed by a company called 454 Life Sciences, which specialized in high-throughput sequencing, based in Branford, Connecticut which was acquired by Roche in 2007. It was commonly referred to as pyrosequencing (Ronaghi et al., 1996). Its chemistry depends on the release of pyrophosphate during the addition of a nucleotide in polymerization which later can be detected using a substrate. It utilizes the abilities of enzyme DNA polymerase followed by ATP sulfurylase and luciferase. Subsequently, apyrase takes care of the remaining nucleotides. Substrates used in these reactions are adenosine 5′ phosphosulfate (APS) and luciferin (Gharizadeh et al., 2007; Nyren and Skarpnack, 2001). Pyrophosphate produced by DNA polymerase during the incorporation of the nucleotide at the 3' end of the growing DNA fragment is converted into ATP by ATP sulfurylase in the presence of substrate APS. This ATP then converts luciferin into oxyluciferin in the presence of enzyme luciferase and thus emits the light, which is proportionate to the amount of pyrophosphate generated. The remaining unused ATP and nucleotides get degraded by apyrase (Mardis, 2013). Another unique feature of this system is the use of a picotiter plate, which helps to miniaturize the system in addition to running many parallel reactions. This results in the production of more than 200,000 reads with an output of 20 Mb per run, where the length of each read was in the range of 100–150 bp in 2005 (Kulski, 2016). With an upgrade in 2008, Roche released a 454 GS FLX Titanium system which enhanced the output to 0.7 Gb data within 24 h, average reads length has been improved up to 700 bp with an accuracy of 99.997%. The platform has shown promising results in detecting real-time base addition as well as for sequencing of inverted terminal repeats (Petri et al., 2014) and secondary structures which used to be difficult to get sequenced (Ronaghi et al., 1999). This platform has displayed two major drawbacks, first with the sequencing of homopolymer regions and the second drawback is prolonged sample preparation for emulsion PCR (Huse et al., 2007; Margulies et al., 2005). Due to the incorporation of a similar nucleotide in the homopolymer region very high signal intensity is produced, which makes it difficult to precisely sequence these regions. Sadly, Roche Diagnostics Corporation has announced the discontinuation of 454 updates in 2013 and a subsequent gradual shut down manufacturing reagents for 454 platforms.

3.3.2 ION TORRENT: PROTON/PERSONAL GENOME MACHINE (PGM) SEQUENCING

Ion Torrent technology was developed by Toumazou et al. in 2006 and Thermo Fisher brought this for commercialization (Rothberg et al., 2011). It utilizes the SBS principle. During the polymerization of the DNA strand, H^+ ions or protons are released, which changes the pH of the microenvironment. This change in the surrounding solution is proportional to the nucleotide added and detected electronically by a semiconductor-based device, which subsequently converts this signal into a sine wave (Merriman et al., 2012). Another feature of this platform is that the polymerization reaction is performed on a CMOS (complementary metal-oxide-semiconductor) chip merged with flow cell and has an electronic sensor at the bottom of each cell. The uniqueness of the Ion Torrent platform lies in the use of unmodified bases which makes it more economical in terms of consumables. Furthermore, the absence of image scans makes it faster for the "base calling" process. Ion torrent can produce reads up to 600 bp (Kulski, 2016). The major drawback of this platform is the interpretation of homopolymer sequences (more than 6 bp) similar to 454 platforms (Kumar et al., 2019; Loman et al., 2012; Morey et al., 2013; Reuter et al., 2015). Since there is no break in the reaction post-base incorporation, due to its chemistry, the addition of the same base produces a similar signal and makes the precise identification of homopolymer sequences very difficult.

3.3.3 ILLUMINA/SOLEXA SEQUENCING

This is the most popular sequencing platform, which was developed by David Klenerman and Shankar Balasubramanian, founders of the Solexa company, later acquired by Illumina (Bentley et al., 2008; Illumina, 2020). The company currently provides benchtop as well as production-scale sequencers. These can generate up to 6TB reads in less than 2 days with 150 bp long paired-end reads. This method is also based on the SBS principle. Here, four different fluorescently modified nucleotides which contain an inactive 3′ hydroxyl group (blocking group) are added to the polymerizing DNA. This blocking group ensures the incorporation of one nucleotide at a time in polymerizing DNA fragments. An optical signal is detected by CCD and post-detection, the fluorescent dye and 3'blocking group are removed so that the next nucleotide can be added in the next cycle (Bentley et al., 2008). This platform is capable of making very short reads as well as producing paired-end short reads. An error rate of this platform is very low and showed 99.9% accuracy. This accuracy is ascribed to the use of blocking group and base-by-base sequencing. Due to this, the accuracy in determining the precise sequence even at the homopolymer region has improved (Bentley et al., 2008; Mardis, 2013; Ross et al., 2013). Despite its superior performance, Illumina sequencing also has some major drawbacks. The first major problem with the Illumina platform is called phasing (Chen et al., 2013). This problem arises when the 3' blocker of nucleotide is not being able to be removed, hence, the next nucleotide fails to incorporate into the growing DNA fragment. This results in the re-detection of the signal from the old nucleotide, which generates a nonsynchronous signal relative to the other growing polynucleotide chains. This fragment will be out of phase or runs one cycle behind the remaining. This asynchronous signal will then affect the quality score of the reads and generates sequences with decreased or compromised quality. Another issue with Illumina sequencing is loading control, as the overloading of the sample may result in a poor quality score in overlapping cluster regions (Dohm et al., 2008; Hutchison, 2007).

3.3.4 ABI/ SUPPORTED OLIGONUCLEOTIDE LIGATION AND DETECTION (SOLID) SEQUENCING

The SOLiD platform was first developed by Jay Shendure et al. and marketed by Life Technologies. It utilizes the principle of SBL (Shendure and Ji, 2008). SOLiD sequencing is also known as polony sequencing as polonies are constructed using a bead-based emulsion PCR with mate-paired DNA fragments in a cell-free system (Kulski, 2016). Sequencing is done using two-base encoding of octamer probe, post-emulsion PCR using primers, which are complementary to the adapters used for library preparation. Sequencing consists of five rounds where each round is 5–7 cycles long. The first cycle begins with the addition of a universal primer complementary to the primer used for sequencing. Then, an octamer probe containing a 5' fluorescent probe is added. This probe is specifically designed where the first two bases are complementary to the bases to be sequence and followed by six degenerated bases which could be complemented to any base (Fig. 3.1).

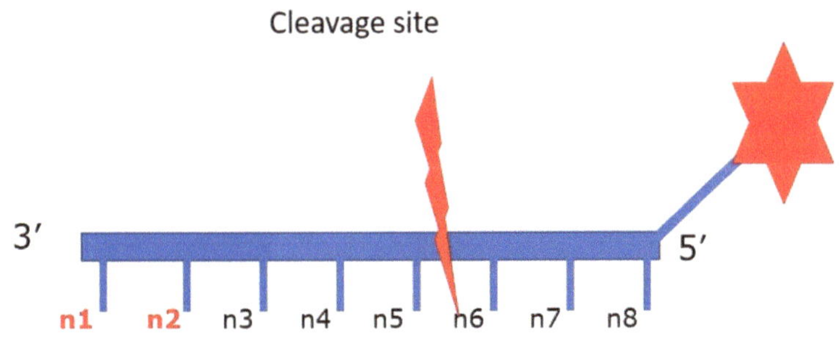

FIGURE 3.1 Octamer probe used in SOLiD sequencing.

As this probe binds, ligation reaction between 5' of the universal primer and 3' of the probe will take place. This probe has a cleavage site between the 5th and 6th base. After ligation, fluorescence is recorded and a probe will be cleaved at the cleavage site. In the next round, the next probe comes and sequences the bases present at base 6th and 7th position for the first base of the previous probe. In the next cycle, the universal primer which is used is complementary to one base below the primer used for emulsion PCR. So that it now covers the first base which was a sequence in the first cycle and this cycle begins with the sequencing of second and third base. The complete reaction sequences each base twice and takes five rounds to complete where it could sequence around 25 bp of DNA fragment from the end base of primer used for emulsion PCR (Clark et al., 2019). Major drawbacks of this method are noise (during ligation), fluorescent leakage, and cross-talk, which can result in misinterpretation of data (Kulski, 2016; Kumar et al., 2019). Another major problem with this sequencing is the very short read length, where the sequencing of a large genome could take up to 14 days. Furthermore, the raw data analysis of this platform requires expertise and a high-end computing machine.

The second-generation sequencing methods have been proved revolutionary in the field of genomics and led to many astonishing scientific achievements; they have failed to address some sequencing applications. To address those gaps, new technologies and platforms of sequencing have emerged in the past decades which are collectively called third-generation sequencing. These technologies are majorly concentrated on single-molecule sequencing and could be grouped in three different groups based on their sequencing approach (Kulski, 2016; van Dijk et al., 2018; Xiao and Zhou, 2020). The first one is SBS-based sequencing method where a single molecule of DNA polymerase is observed for synthesizing the single DNA molecule (Rhoads and Au, 2015). The second approach utilizes nanopore, where bases of single DNA molecules are detected as they passes through that nanopore (Wang et al., 2014). The third approach uses the power of advanced microscopy, utilizing the scanning tunneling electron microscopy (TEM) and fluorescence resonance energy transfer (FRET), and finally detects the sequence of a single DNA molecule (Kulski, 2016).

Since these technologies examine a single DNA molecule, the issues related to PCR-amplification introduced biases, and dephasing is solved. These technologies offer more accurate, faster, and more economical sequencing.

Major third-generation platforms are discussed in detail below:

3.3.5 SINGLE-MOLECULE REAL-TIME (SMRT) SEQUENCING

SMRT platform is marketed by Pacific Biosciences. It utilizes the principle of SBS (Kulski, 2016; Kumar et al., 2019; Rhoads and Au, 2015; van Dijk et al., 2018; Xiao and Zhou, 2020). Here, template DNA is called SMRT bell, a closed circular DNA molecule, which is prepared by ligating the hairpin adapter to both ends of double-stranded DNA. This technique utilizes the power of a zero-mode waveguide (ZMW) to real-time sequence the single DNA molecule. These ZMWs are present on SMRT cells. One SMRT cell contains up to 150,000 ZMWs. At the bottom of each ZMWs, a single molecule of DNA polymerase is immobilized. Four nucleotides which are tagged with distinct fluorophores were then introduced to the SMRT cell. When a nucleotide passes the bottom of ZMW, a fluorescence signal is detected. If that nucleotide is incorporated into the growing strand, the signal will be longer, whereas if the nucleotide is not incorporated and just passed, the signal is short. After nucleotide incorporation, the polymerase cuts the 5' fluorophore and translocates it to the next base. These fluorescence signals are then interpreted into a base sequence called continuous long reads (CLR). Since the template DNA or SMRTbell is a closed circle, the DNA polymerase can sequence both the strands and depending upon the half-life of DNA polymerase, it sequences both the strands multiple times. The biggest advantage of SMRT sequencing is the read length. It has been demonstrated by PacBio that the PacBio RS II system with C4 chemistry could produce a read length beyond 10 kb. Despite these advantages, SMRT sequencing has shown two major disadvantages; lower throughput and higher error rate. Due to failure in attaching DNA polymerase or incorporation of more than one DNA molecule, out of 150,000 ZMWs on a SMRT cell, only 35,000–70,000 ZMWs generate successful reads.

3.3.6 NANOPORE SEQUENCING

Nanopore sequencing was first principally demonstrated by Deamer et al. in which α-hemolysin nanopore is used to detect DNA translocation (Kasianowicz and Bezrukov, 2016; Wang et al., 2014). This platform was commercialized by Oxford Nanopore Technologies. This technique utilizes the principle that when a specific nucleotide base is passed across a nanopore, then the ion current in that pore is changed which is specific to that nucleotide (Kono and Arakawa, 2019). In this sequencing, one strand of DNA molecules is being attached to a leader molecule and motor protein. There are two chemistries used -1D, only one strand is sequenced, 2D, a hairpin adapter is used, and both the strands are synthesized to prepare the DNA library. Motor protein guides the leader molecule through the pore, followed by the template strand, hairpin adapter, and complementary strand (if 2D chemistry is used). Motor protein unwinds the ds DNA. The DNA is passed at a speed where only one base has crossed the nanopore at a time. This passing of DNA creates a deflection in current which is passed across the pore, and changes in the current are specific to the base present in the DNA fragment. This change in the current is recorded by the sensor present on the flow cell and used for base calling. The major drawbacks of these technologies include the lower throughput and high error rate (Gupta and Gupta, 2014).

3.3.7 TRUE SINGLE-MOLECULE SEQUENCING

True single-molecule sequencing (tSMS) was marketed by Helicose Biosciences, also known as Helicose sequencing (Hart et al., 2010). Post-bankruptcy of Helicose Biosciences, this service is provided by SeqLL Inc (Kulski, 2016). The basic principle of tSMS is also the SBS. The fragmented DNA molecules are unzipped into single-stranded and a poly-A tail is added to the 3' end (Hart et al., 2010). The last dATP used in this is fluorescently labeled. This poly-A-tailed ssDNA is then introduced on Helicose-flow cells. The surface of these glass flow cells is coated with poly-T which are then hybridized with poly-A tail of DNA fragment and loaded on the sequencer, where the CCD camera locates the sequences with the fluorescently labeled dATP. Post-imaging, the fluorescent tag is cleaved. Four dNTPs that are fluorescently tagged were added in sequence. After each addition, images are taken by CCD camera, followed by slicing of the fluorescent tag. These images are then used for base calling. The biggest advantage of this technique is the use of a single molecule that could be sequenced without PCR, thus avoiding amplification-induced errors. Furthermore, this technique could sequence RNA directly from a single cell without reverse-transcription. Also, it is economical. Higher raw error rate due to single-molecule sequencing and shorter read length are the major disadvantages of this method (Gupta and Gupta, 2014).

3.4 APPLICATION OF OMICS APPROACH IN PLANT MOLECULAR ECOLOGY

Consideration of the use of NGS of nonmodel species in ecological context raises the question—how far genomes can affect the processes at the community and ecosystem level? It is still a matter of debate. Population studies about forest trees such as oak (Madritch and Hunter, 2002), *Metrosideros polymorpha* (Treseder and Vitousek, 2001), and *Populus* (Schweitzer et al., 2004) have shown that genetic variations among populations have strong and immediate effects on nutrient cycling in ecosystems because of the strong relation of litter chemistry to decomposition. In this context, diversity of microorganisms in soil may show possible dependence on the diversity of trees. Such examples have been shown in studies regarding the diversity and assemblage of tree-dwelling arthropod community which is affected by the genetic diversity of *Populus* (Wimp et al., 2005), *Solidago altissima* (Crutsinger et al., 2006), and *Oenothera biennis* (Johnson et al., 2006). Interdependence of this organismic diversity affects the net productivity positively at the level of the ecosystem as plants are the primary producers and soil fertility depends upon an efficient nutrient cycle. These kinds of studies constantly prove the significance of efforts to understand the interactions among genomes at a higher level of an organization like "community genome" (Ungerer et al., 2008) and how they function. The implication of NGS in plant molecular ecology can be divided into the following three components:

3.4.1 APPROACHES TO ESTIMATE GENETIC VARIATIONS

3.4.1.1 GENOMIC VARIATIONS; MICROSATELLITE SCANNING

Traditionally, neutral markers, such as simple sequence repeats (SSRs) and AFLPs (amplified fragment length polymorphisms) have been used by ecologists to detect gene flow, inbreeding, genetic drift, and demography. In one NGS run, depending upon the chemistry used, around a few gigabase data can be obtained. These bases are mainly part of short reads originating from the ends of shotgun fragments. These reads can be assembled in large contigs with the help of bioinformatics tools available. A couple of softwares have been developed to scan the NGS data for searching microsatellite loci throughout the genome. Among the available programs for microsatellite search SciRoKo is a fast search algorithm with a built-in statistical summary tool (Kofler et al., 2007). Further, MsatCommander has been introduced with a graphic user interface which is a platform-independent program (Faircloth, 2008). Additionally, QDD is a similar program with added facility of elimination of adapters and redundant sequences, tag sorting, and detection of possible PCR (Polymerase Chain Reaction) duplicates (Meglecz et al., 2010). An improved tool High SSR provides the faster approach by eliminating the need for all-against-all-alignment and therefore allows the development of mappable SSR markers (Churbanov et al., 2012).

However, these neutral markers do not provide an idea about the effect of genetic variation on selection pressure because all these markers based on the repeat regions lie within the noncoding DNA sequences which lack genes. Moreover, due to lack of knowledge about the function of noncoding DNA, these markers do not allow us to interpret the effect on evolution. As the genome sequences of more organisms are provided, genomic scans are facilitated to differentiate between neutral and nonneutral markers. Genomic approaches enable ecologists to study if the genes with specific functions are differentially expressed under certain conditions, that is, in response to inbreeding or outbreeding.

Next-generation sequencing approaches involving genomic sequences facilitate SNP profiling, which gives an overall estimate of genomic variations. These genomic variations can be used for association studies, which might indicate the genetic mechanisms of plants and their response to the environment.

3.4.1.2 STUDY OF VARIATIONS USING SNP PROFILING

Genomic reads obtained through next-generation sequencing provide a wide array of single nucleotide polymorphisms (SNPs). Nucleotide variation between individuals can be easily identified using the reads obtained from NGS sequencing which is represented by a single nucleotide variation among reads at a particular position in the genome. Thousands of SNPs can be identified through a single NGS run. GigaBayes (Marth et al., 1999) is one of the oldest approaches to detect single nucleotide variations throughout the genome. Advanced tools like SNP detectors (Zhang et al., 2005) are efficient softwares to precisely detect inter-species and intra-subspecies variations. Another useful pipeline inGAP (integrative next-generation genome analysis pipeline) takes the Bayesian principle into account (Qi et al., 2010) and suits to detect SNPs and insertions/deletions (indels). Additionally, an open-source tool VarScan is available through the internet for the detection of variants. It is compatible with different aligners too (Koboldt et al., 2009). Another web-based tool is SNiPlay (Dereeper et al., 2011), which integrates with other pipelines like haplotype reconstruction and linkage disequilibrium evaluation. In this line, QualitySNPng (Nijveen et al., 2013) is a user-friendly tool to detect as well as an interactive visualization of the SNPs. It uses a haplotype-based strategy and does not require a reference genome sequence for identifying reliable SNPs using RNAseq and genomic sequencing data. It provides a user-friendly interface for multiple filtering and handling sequencing errors. Verification of SNPs obtained from sequence analysis is essential to avoid false-positives. High coverage depth is one of the requirements for this purpose which is possible with NGS techniques. Increased coverage with a large number of samples might prove costly; on the other hand, SNPs detected in neutral regions give less information in terms of adaptive and selective processes. To rationalize the efforts of detecting reliable SNPs, two strategies can be acquired, first is to sequence the transcriptome which is around 2–5% of the total genome and the second is to sequence RAD tags (Restriction site-Associated DNA tags) (Miller et al., 2007) which permits high coverage depth of reads cost-effectively. Subsequent validation of the variants, such as insertions–deletions (indels) and copy number variation (CNV) can be done using SMRT sequencing technologies, such as the Pacific Biosciences (PacBio) and Oxford Nanopore systems (Goodwin et al., 2016; Lee et al., 2016).

3.4.1.3 GENOME SKIMMING AND PLASTID GENOMICS ASSISTS PHYLOGEOGRAPHY; FROM POPULATION GENETICS TO POPULATION GENOMICS

The genome skimming approach, which provides a shallow sequence of an organism, proves beneficial in plant molecular ecology where a complete chloroplast genome and its comparative analyses are used to develop molecular markers using divergence hotspot (L. Liu et al., 2020). Chloroplast simple sequenced repeats (cpSSRs) and genomic simple sequenced repeats (gSSRs) developed through the genome skimming approach are useful for developing plastid and nuclear phylogenies, respectively. Comparison of these two phylogenies in Saxifragaceae revealed an important evolutionary process of plastid capture indicating ancient hybridization through backcrossing (L. X. Liu et al., 2020). DNA sequence from chloroplast which is maternally inherited without any recombination can solve the issues related to phylogeographic patterns and population dynamics. Comparative genomics of the chloroplast genomes of *Quercus rubra* and *Castanea mollissima* B. have been successfully applied to barcode Chinese Oak species with the help of ITS region of 5.8 sRNA and a single copy nuclear gene (Yang et al., 2017).

3.4.1.4 NONNEUTRAL MARKERS DEVELOPED USING THE AVAILABLE GENOME SEQUENCES

Despite the wide use for phylogeographic and population dynamics studies, organellar genome sequences are less useful for studying ecological interactions because of uniparental inheritance and lack of recombination. Phylogeneticists often use markers for tracing the evolution of particular nuclear genes as these carry variations that provide the platform for selection pressure. These markers include NPCLs (nuclear protein coding loci), EPIC (exon-primed intron-crossing), and ANMs (anonymous nuclear markers). Homology detection is a consensus method to design these markers. Already annotated genomic databases provide a more confident design of these markers (Thomson et al., 2010). Through alignment of transcriptome reads, obtained from two distantly related nonmodel organisms to the reference genome of the model species, consensus sequence regions are obtained which can be used to identify short-conserved regions from low-copy expressed genes serving as NPCLs and EPICs. Similar techniques have been applied using RNAseq reads of *Cardamine hirsuta* and *Arabis alpina* and thousands of nuclear markers have been developed (Stockenhuber et al., 2015). A recent phylogenetic study in *Medicago* using next-generation sequencing data has successfully explained the incongruence of earlier phylogenies (de Sousa et al., 2014). Transcriptomic studies prove useful in many aspects. First, it is helpful to differentiate between neutral and nonneutral, second, it provides the quantitation of the gene activity in addition to the sequence variation.

3.4.2 ASSOCIATION STUDIES; GENETIC VARIATIONS RELEVANT TO THE PHENOTYPE

3.4.2.1 GENOME-WIDE ASSOCIATION STUDY

Cost-effective sequencing methods have made it possible to sequence more individuals to identify genetic variations although assigning these genetic variations to a relevant phenotype gives insight into the ecological processes related to the particular variation. Genome-Wide Association Study (GWAS) approaches are useful to assign particular polymorphisms to a responsive phenotype in plants (Anderson et al., 2019; Kress and Erickson, 2007). Globally, climate change is assumed to be responsible for the emergence of new ecological agents, which expose hosts to new pathogen reservoirs. Considering this, identification of genetic variants relevant for the natural variation in resistance as well as genetic determinants of pathogens for disease emergence is necessary to measure the adaptive potential of the wild populations. Candidate SNPs in the association of long-term quantitative data on growth and survival are useful to pinpoint the putative genes under selective pressure in the process of physiological adaptation to climate variations and their consequences. GWAS in conjuncture with next-generation sequencing proves to be a powerful tool, to predict the genes responsible for adaptive resistance toward pathogens in wild populations. Traditional linkage mapping with markers associated give the information about the linkage disequilibrium regions of the recombinant inbred lines (RIL) and their segregated progenies, but GWA mapping with NGS techniques allows efficient mapping of resistance-related genetic variations in natural populations accumulating recombinant events over thousands of generations (Bartoli and Roux, 2017). A GWA study based on 23 haploid *Heterobasidion annosum* collected from different European countries has found 12 SNPs, significantly associated with the virulence of this necrotrophic fungal pathogen which is responsible for severe damages in forest conifers (Dalman et al., 2013). The combination of these approaches opens a recently emerging strategy for the estimation of heritability of plant–pathogen interactions using association of phenotypic and genotypic variation of both the biotic partners (Roux and Bergelson, 2016). CRaVe, a software package that is open source and freely available through the internet which is designed to perform a range of association tests between genotype (sets of SNPs) and a phenotype using Unix or R (Ferguson et al., 2013).

3.4.2.2 GENOME-WIDE SELECTION SCAN

Next-generation sequencing data enables genome-wide selection scan (GWSS) for the mutations which get selected or purified from the organismic genome. A similar study of a population of spruce trees in a forest ecosystem taking two species *Picea glauca* and *Picea engelmannii* and their hybrid crosses revealed about 68 loci under putative selection pressure in response to growth and phenology, insect herbivory resistance as well as cold hardiness (De La Torre et al., 2014). GWSS Study in 44 natural accessions of *Brachypodium distachyon* highlights the impact of environmental constraints on natural selection (Bourgeois et al., 2018). The use of software like SweepFinder 2 (Collevatti et al., 2019) has efficiently revealed the selective sweeps and genetic hitchhiking events by analyzing SNP data obtained from the whole-genome sequence of 128 individuals across 13 populations of a Neotropic tree, *Handroanthus impetiginosus* (Huber et al., 2016).

3.4.2.3 GENE-ENVIRONMENT ASSOCIATIONS STUDIES

Gene-environment associations studies (GEAS) are important, as environment and genetic variations are related through the platform of natural selection. The genotypes, which result in the phenotypes that are fit for the survival of the individual in a particular environment, are selected through natural selection. Only, these genotypes are available to recombine with each other, the rest of the unfit genotypes die and get eliminated. In this context, GEAS gives a complete representation of the interaction of genomes with ecological impacts in the course of a longer segment of time, which includes different alterations in climate and biota. GEAS play an important role in the investigation of biological pathways involving the genes which function in a specific environment or in response to external exposures that are hazardous to susceptible individuals. Such a kind of genomic and transcriptomic approach is useful for identifying marker patterns for ecosystem hazards and accordingly safety threshold can be predicted (Thomas, 2010). Recently emerging fields of "landscape genomics" (Rellstab et al., 2015) and "restoration genomics" (Carvalho et al., 2020) are supported by NGS data and their outcomes, that is, whole-genome variations.

3.4.3 IMPLICATIONS OF WHOLE-GENOME SEQUENCE DATA IN PLANT MOLECULAR ECOLOGY

3.4.3.1 CONTRIBUTION OF NEXT-GENERATION SEQUENCING IN CONSERVATION BIOLOGY; INBREEDING VS. OUTBREEDING DEPRESSION

In conservation biology, it is normally observed that inbreeding depressions resulting from a small size of the self-compatible population are hazardous, although sometimes introducing out members from a long-distance might prove alarming. In a study on *Stylidium hispidum,* it has been shown that an intermediate optimal outcrossing distance in the species is important because of restoration (Hufford et al., 2012). In this matter, NGS is very useful to assess the suitability of introgression of diverse genotypes to preserve biodiversity and retain adaptation to the environment.

Conservation of endangered species is a mammoth and useful task in ecology. However, it has to be predetermined whether the endangered species is in conservation priority. In one such example explained further, putative endangered species proved to be an occasional hybrid of the two most commonly occurring species. Interspecies comparative population genomics of the three species of genus *Rhododendron* resulted in the conclusion that the endangered species *Rh. pubicostatum* (Subsect. *Teliensia*) is a hybrid of *Rh. bureavii* and *Rh. sikan-gensevar.exquisitum* (Zhang et al., 2020). This study comparing the parent species with hybrid using 12 microsatellite markers developed through NGS of 72 individuals of population indicated that targeted plant harbors higher biodiversity than a parent with frequent introgression of the latter.

3.4.3.2 STUDY OF ADAPTATION AND INVASION

Adaptation of different species to a similar environment is one question, which can be answered using the next-generation sequencing approach. Transcriptome characterization of two mangrove species *Rhizophora mangle* (Rhizophoraceae) and *Heritiera littoralis* (Malvaceae) has been done and obtained data were aligned to already annotated genomes of *Arabidopsis* and *Populus*. One exhibited more similarity to *Arabidopsis* and another to *Populus* which clearly showed the divergent evolution of two species. However, both mangrove species have shown substantial correlations in gene expression profiles, which deviated significantly from the model plant species (Dassanayake et al., 2009). This study concludes that similar genetic mechanisms dominate in plants descending from different lineages but facing similar extreme environments.

Sometimes genome-wide studies in an ecological context reveal important molecular and biochemical phenomenon regarding the invasion. The accidental introduction of knapweed from Europe to Northern America has outcompeted native plants because of the allelochemical catechin. The whole-genome transcriptome study on the model plant *Arabidopsis thaliana* grown with knapweed revealed a wave of reactive oxygen species which triggers calcium signaling cascade and thus causes genome-wide changes in transcriptome, leading to root system death in affected species (Bais et al., 2003).

3.4.3.3 IMPORTANT PHYLOGENETIC EVENTS

The evolution of advanced characters in modern-day plants is the outcome of ecological interactions at the previous time and place. Separate genetic studies reveal the molecular basis of the evolution of floral symmetry, corolla form, flower color, and scent as well as inflorescence (Chen et al., 2017; McCarthy et al., 2017; Spigler, 2017). Zygomorphic symmetry of flower has been found to have been evolved independently in separate domains. However, the replacement of genetic studies with genomics can efficiently elucidate the possible role of natural selection of ecologically important interactions. Sequencing the data of nucleotides in conjuncture with the protein family information Pfam (El-Gebali et al., 2019) has revealed complex relationships between duplication mechanisms and variations in the gene family with turnover (Wang et al., 2018).

3.4.3.4 EVOLUTION OF SPECIFIC GENETIC MECHANISMS; CANDIDATE APPROACH

Candidate approaches mainly rely upon sequence-based homology. In this era, when huge sequencing data are available, searching for homology of ecologically relevant genes is not a difficult task. Genome-wide comparisons at proteome level in conjuncture with the information of protein-coding genes, expression data, and promoter analyses of arbuscular mycorrhizae (AM)-competent host species with nonmycorrhizal species, successfully revealed candidate genes and possible genetic mechanisms responsible for symbiosis (Favre et al., 2014). Focused studies about these will provide important ecological perspectives about symbiotic plant–microbe interactions.

3.4.3.5 ENVIRONMENT SAMPLING; EFFECT OF POLLUTION

Understanding the interaction of genomes within a higher level of organization is challenging, however, metagenomics of microbial communities have been proven successful. As a refinement of nucleic acid isolation technique from complex samples has evolved, sequencing of environment samples directly facilitated the identification of new metabolic pathways within the microbial community. These can be soil samples or aquatic samples.

This approach can reveal the novel interaction in some microbial consortia. Sequencing of environmental samples of biofilms from acid mine drainage in a study paves the way to determine nearly complete genomes of dominant bacteria. Further analyses revealed the presence of a small population of *Leptospirillium* group III bacteria, which contained genes for nitrogen fixation (Tyson et al., 2004). This species of bacteria happened to be a key species of the community owing to its ability to fix nitrogen in an environment without external nitrogen input.

A similar approach has proven useful to identify beneficial processes. A novel phototrophy knew as photoorganotrophy exhibited in marine bacterioplankton. The new rhodopsin with photochemical reaction cycle intermediates and kinetics of archaeal rhodopsin activates an alternative way to use energy from the sun in conditions limiting for the organic matter that suggests differential pathways of carbon cycle on the surface and deep-sea (Beja et al., 2000). Additionally, a pilot study conducted on the Sargasso Sea near Bermuda generated around 1.045 billion base pairs of nonredundant sequences. Over 1.2 million previously unknown genes, out of which 70,000 were novel and some of which function in a wide range of biogeochemical pathways (Venter et al., 2004).

3.4.3.6 PHYLOGENETICS; DN/DS RATIO

Before the NGS era, a study of genetic drift and inbreeding had been carried out using traditional markers. Although due to the unavailability of sequence details, the effect of selection pressure working upon those genetic changes was difficult to predict. The availability of ample sequencing data has made it possible to trace the signatures of selection pressure and deduce the possible interspecific and intergeneric interactions. Natural selection is an inevitable component of an ecosystem and selection pressure works on the changes occurred in the genome. Readily available next-generation sequencing data can be explored to specify the regions of selection sweeps and signatures of recent positive selection. CoGe (Comparative Genomics) database (Grover et al., 2017; Lyons et al., 2008; Nelson et al., 2018) is a useful databank of completed and nearly completed plant genomes. Retrieved sequences from synteny-based searches can be analyzed for the completely annotated sequence on MEGANTE which is a web-based program for integrated plant genome annotation (Numa and Itoh, 2014). Whether genetic variations, introduced through mutations, would be translated into altered response toward the biotic and abiotic interactions of the organism or not, depends upon the nature of the mutation if that is synonymous or nonsynonymous and linked to the coding region to be translated. These random mutations get selected or purified during interactions of the organism with the environment. The ratio of nonsynonymous and synonymous mutations (dN/dS) coupled with maximum likelihood and Bayesian approach indicates toward the signatures of positive selection which reveals the regions of ecological importance and novel functional relations. PAML (phylogenetic analyses by maximum likelihood) is a useful and sensitive tool for this approach (Yang, 1997, 2007). NGS databases can be used to mine the information, which can be analyzed to identify altered processes that are preferably selected as a consequence of successful ecological interactions. In case of unavailability of the sequence, data can be generated on NGS platforms as done in a study on *Eucalyptus grandis* which revealed the purifying selection throughout the genome (Novaes et al., 2008). Genomics tools can also be developed through de novo transcriptome assembly of ecologically important nonmodel organisms as done in the case of *Scabiosa columbaria* (Angeloni et al., 2011).

3.4.3.7 PHENOTYPIC VARIATION/PLASTICITY DUE TO EPIGENETIC COMPONENT

Increasing resources of next-generation sequencing facilitates the comparison of genomes and associations of phenotypic variations with genotypic coding. The persistence of populations in changing climatic conditions relies on two strategies, the first is local adaptations in which specific genotypes are favored, and the second relates to the existing genetic diversity, which produces new phenotypes to fit for the changing environment. Plants are sessile organisms, which rely on adaptation toward environmental conditions more than any other organism on the planet. Besides, the invasiveness of a species depends upon its capacity to adapt to new environmental conditions deciphering the involvement of high genetic variability. However, invasion of the clonal population involves severe bottlenecks. Despite these challenges, seagrass species *Zostera marina* thrive well in a clonal meadow because of methylation variance, independent of genetic variation (Jueterbock et al., 2020). Similarly, *Alternanthera philoxeroides* (alligator weed) is highly invasive, which invades through asexual reproduction across china. Transcriptomics reveals that invading toward low-temperature zone is supported by altered gene expression of the protein kinases, transcription factors, metabolite biosynthetic pathways genes, plant hormone response genes, and genes involved in plant–pathogen interaction, which provide the cold tolerance to the weed plant (Luo et al., 2020). Further, the involvement of H3K9Ac and H3K4Me3 is responsible for the differential expression of the above-mentioned genes. Next-generation sequencing analyses like BS-RADseq (BiSulfite-Restriction Site-Associated DNA tags sequencing) can be used to assess the methylome through which epigenetic variation can be related to phenotypic variations (Trucchi et al., 2016). Studies associating heritable epigenetic modifications in *Hydrocotyle vulgaris* showed that genetic variations did not correlate with Intra-/inter-population genetic variation and phenotypic variations. However, epigenetic diversity was correlated with phenotypic variations like petiole length and specific leaf area (Wang et al., 2020).

3.4.3.8 GENETIC DRIFT; DETECTION THROUGH GENOMICS

Plant pathogens, mainly RNA viruses, impose important ecological challenges. However, during transmission and within-host movement of viral populations, due to small population size, stochastic changes may occur in population structure leading to the genetic drift (Zwart and Elena, 2015). Fast replication with less fidelity generates mutant swarms (Andino and Domingo, 2015) in the viral population which might recombine at a high rate within the host (Froissart et al., 2005; Tromas et al., 2014). This diversity enables the large populations to quickly adapt to changing conditions as competition between haplotypes (generated through the mutation) with different fitness results in selection pressure (Elena et al., 2014). A similar study involving deep sequencing of small RNAs of time-sampled population of potato virus Y revealed signatures of narrow genetic bottlenecks and strong selective sweeps followed by genetic drift of selected LD (linkage disequilibrium) blocks (Kutnjak et al., 2017). This study also detected strong diversifying evolution of the virus in the host which is efficient in the accumulation of salicylic acid thus being less susceptible. Therefore, the level of diversity within the host population also plays important role in the invasion and adaptation of the pathogen species.

3.4.3.9 ENVIRONMENTAL STRESS AND CHANGES IN GENOMIC PATHWAYS

Plant populations are different than animals as they cannot move when there is environmental stress. Therefore, in response to the introduction of stress in a landscape, they need to adapt with fast acclimation strategies. Epigenetic changes are one of the swift changes which occur as short-term adaptive responses. Transgenerational heritability of epigenome and environmental linkage of phenotypic plasticity in fitness-related traits may add important insight to adaptive variations (Whipple and Holeski, 2016). The epigenetic changes are capable of modifying important response regulators which in turn might switch to different genetic pathways to cope up with extreme environmental conditions. It is interesting to investigate how adaptive transgenerational plasticity and so-called "stress memory" get inherited (Weinhold, 2018).

3.5 LIMITATIONS

"Ecological genomics" (Gilad et al., 2009) is indeed emerging as a multidisciplinary field. Research in molecular ecology requires different areas of expertise and additionally conceptual understanding and integration of mechanistic interactions at different levels is important. Although transferring the genomic information from model organisms to nonmodel ones solves some questions. Expansion of sequencing approach to these nonmodel species amenable to study gives the field more strength. Whole-Genome Sequencing projects running in different research centers have sequenced 1000 plant reference genomes which have enriched the database (Wang et al., 2020). Furthermore, functional tests of genetic outcomes of the studies are also required for which new methodologies have to be developed. Sometimes genomic scans might fail to map local adaptation genes because of the coupling of intrinsic pre-zygotic or post-zygotic incompatibilities which create an endogenous barrier to gene flow and exogenous barriers like environmental constraints (Bierne et al., 2011).

Choosing the appropriate statistical method and sampling size is the key to a successful study in plant Ecogenomics. Sequencing of nonmodel plants with the sampling satisfying power statistically might prove costlier for an ecological study. One serious issue with the costeffective sequencing approach is that when reduced representation technologies are used, the problem of missing data arises. Another problem with next-generation sequencing is duplicated regions. In plants, events of duplication of genomes and increase in ploidy level are frequent. Analyses pipelines, available for data obtained from sequencing by synthesis, do not provide a better way to handle repeated sequences (McKain et al., 2018).

The development of functional tests to validate the outcome of sequencing data analyses for a large number of individuals is a challenging task. Improving the quality and quantity of the NGS data cost-effectively is the next challenge to deal with. Data storage and compaction algorithms compatible with DNA sequence information are needed to be worked upon.

3.6 FUTURE PROSPECTS

NGS technique is making our database richer day by day, in terms of DNA sequence information. Analyses of available sequence information in the nonmodel organism might reveal the interaction of genetic variations with environmental conditions and transgenerational inheritance of selectively important variations. Scanning genetic variations in a spatiotemporal manner is important. Through the whole-genome approach, it is possible to differentiate between neutral and nonneutral markers. Scanning these markers within populations and across the generations gives an idea about markers of selective importance as well as genetic drift in population. This approach can be implicated successfully in conservation biology. Genome X Environment studies reveal the genetic effects of climate change, habitat fragmentation, and environmental degradation. These might facilitate setting up the alarming threshold to save the habitat. The whole-genome approach enables sampling of many individuals and genotyping of many markers, which differentiates locus-specific effects, such as mutation, recombination, and selection from genome-wide effects like gene flow, inbreeding, and genetic drift (Ouborg et al., 2010). A promising future for genome-wide studies of nonmodel species in ecological context might be seen in the efforts of several platforms which provide compatible tools to analyze data and opportunity to collaborate for the genome studies as well as sequencing new genomes, annotations of genome assemblies, and to curate a different kind of data, sequences, and studies associated with the specific organismic genome.

KEYWORDS

- **ecogenomics**
- **conservation**
- **next-generation sequencing**
- **adaptation**
- **invasion**
- **evolution**

REFERENCES

Anderson, R.; Edwards, D.; Batley, J.; Bayer, P. E. Genome-Wide Association Studies in Plants [https://doi.org/10.1002/9780470015902.a0027950]. *eLS* **2019,** 1–7. https://doi.org/https://doi.org/10.1002/9780470015902.a0027950 (Major Reference Works).

Andino, R.; Domingo, E. Viral Quasispecies. *Virology* **2015,** *479–480*, 46–51. https://doi.org/10.1016/j.virol.2015.03.022

Angeloni, F.; Wagemaker, C. A.; Jetten, M. S.; Op den Camp, H. J.; Janssen-Megens, E. M.; Francoijs, K. J.; Stunnenberg, H. G.; Ouborg, N. J. De novo Transcriptome Characterization and Development of Genomic Tools for Scabiosa Columbaria L. Using Next-Generation Sequencing Techniques. *Mol. Ecol. Resour.* **2011,** *11* (4), 662–674. https://doi.org/10.1111/j.1755-0998.2011.02990.x

Attwood, T. K. Concepts, Historical Milestones and the Central Place of Bioinformatics in Modern Biology: A European Perspective, Bioinformatics—Trends and Methodologies, Mahmood A. Mahdavi. *IntechOpen* **2011,** Ch. 1. https://doi.org/- 10.5772/23535

Bais, H. P.; Vepachedu, R.; Gilroy, S.; Callaway, R. M.; Vivanco, J. M. Allelopathy and Exotic Plant Invasion: From Molecules and Genes to Species Interactions. *Science* **2003,** *301* (5638), 1377–1380. https://doi.org/10.1126/science.1083245

Bartoli, C.; Roux, F. Genome-Wide Association Studies In Plant Pathosystems: Toward an Ecological Genomics Approach [Review]. *Front. Plant Sci.* **2017,** *8* (763). https://doi.org/10.3389/fpls.2017.00763

Beja, O.; Aravind, L.; Koonin, E. V.; Suzuki, M. T.; Hadd, A.; Nguyen, L. P.; Jovanovich, S. B.; Gates, C. M.; Feldman, R. A.; Spudich, J. L.; Spudich, E. N.; DeLong, E. F. Bacterial Rhodopsin: Evidence for a New Type of Phototrophy in the Sea. *Science* **2000,** *289* (5486), 1902–1906. https://doi.org/10.1126/science.289.5486.1902

Bentley et al., Accurate Whole Human Genome Sequencing Using Reversible Terminator Chemistry. *Nature* **2008,** *456* (7218), 53–59. https://doi.org/10.1038/nature07517

Bierne, N.; Welch, J.; Loire, E.; Bonhomme, F.; David, P. The Coupling Hypothesis: Why Genome Scans May Fail to Map Local Adaptation Genes. *Mol. Ecol.* **2011,** *20* (10), 2044–2072. https://doi.org/10.1111/j.1365-294X.2011.05080.x

Bourgeois, Y.; Stritt, C.; Walser, J.-C.; Gordon, S. P.; Vogel, J. P.; Roulin, A. C. Genome-wide Scans of Selection Highlight the Impact of Biotic and Abiotic Constraints in Natural Populations of the Model Grass *Brachypodium distachyon* [https://doi.org/10.1111/tpj.14042]. *Plant J.* **2018,** *96* (2), 438–451. https://doi.org/https://doi.org/10.1111/tpj.14042

Burleigh, S. H.; Cavagnaro, T.; Jakobsen, I. Functional Diversity of Arbuscular Mycorrhizas Extends to the Expression of Plant Genes Involved in P Nutrition. *J. Exp. Bot.* **2002,** *53* (374), 1593–1601. https://doi.org/10.1093/jxb/erf013

Carvalho, C. S.; Forester, B. R.; Mitre, S. K.; Alves, R.; Imperatriz-Fonseca, V. L.; Ramos, S. J.; Resende-Moreira, L. C.; Siqueira, J. O.; Trevelin, L. C.; Caldeira, C. F.; Gastauer, M.; Jaffe, R. Combining Genotype, Phenotype, and Environmental Data to Delineate Site-Adjusted Provenance Strategies for Ecological Restoration. *Mol. Ecol. Resour.* **2020**. https://doi.org/10.1111/1755-0998.13191

Chen, F.; Dong, M.; Ge, M.; Zhu, L.; Ren, L.; Liu, G.; Mu, R. The History and Advances of Reversible Terminators Used in New Generations of Sequencing Technology. *Genom. Proteom. Bioinfo.* **2013,** *11* (1), 34–40. https://doi.org/10.1016/j.gpb.2013.01.003

Chen, L.; Zhang, B.; Li, Q. Pollinator-mediated Selection on Flowering Phenology and Floral Display in a Distylous Herb *Primula alpicola*. *Sci. Rep.* **2017,** *7* (1), 13157. https://doi.org/10.1038/s41598-017-13340-0

Churbanov, A.; Ryan, R.; Hasan, N.; Bailey, D.; Chen, H.; Milligan, B.; Houde, P. HighSSR: High-throughput SSR Characterization and Locus Development from Next-gen

Sequencing Data. *Bioinformatics* **2012,** *28* (21), 2797–2803. https://doi.org/10.1093/bioinformatics/bts524
Clark, D. P.; Pazdernik, N. J.; McGehee, M. R. DNA Sequencing. *Mol. Biol.* **2019,** 240–269. https://doi.org/10.1016/b978-0-12-813288-3.00008-2
Collevatti, R. G.; Novaes, E.; Silva-Junior, O. B.; Vieira, L. D.; Lima-Ribeiro, M. S.; Grattapaglia, D. A Genome-wide Scan Shows Evidence for Local Adaptation in a Widespread Keystone Neotropical Forest Tree. *Heredity* **2019,** *123* (2), 117–137. https://doi.org/10.1038/s41437-019-0188-0
Crutsinger, G. M.; Collins, M. D.; Fordyce, J. A.; Gompert, Z.; Nice, C. C.; Sanders, N. J. Plant Genotypic Diversity Predicts Community Structure and Governs an Ecosystem Process. *Science* **2006,** *313* (5789), 966–968. https://doi.org/10.1126/science.1128326
Dalman, K.; Himmelstrand, K.; Olson, A.; Lind, M.; Brandstrom-Durling, M.; Stenlid, J. A Genome-wide Association Study Identifies Genomic Regions for Virulence in the Non-model Organism *Heterobasidion annosum* s.s. *PLoS One* **2013,** *8* (1), e53525. https://doi.org/10.1371/journal.pone.0053525
Dassanayake, M.; Haas, J. S.; Bohnert, H. J.; Cheeseman, J. M. Shedding Light on an Extremophile Lifestyle through Transcriptomics. *New Phytol.* **2009,** *183* (3), 764–775. https://doi.org/10.1111/j.1469-8137.2009.02913.x
De La Torre, A. R.; Wang, T.; Jaquish, B.; Aitken, S. N. Adaptation and Exogenous Selection in a Picea glauca x Picea engelmannii Hybrid Zone: Implications for Forest Management under Climate Change. *New Phytol.* **2014,** *201* (2), 687–699. https://doi.org/10.1111/nph.12540
de Sousa, F.; Bertrand, Y. J.; Nylinder, S.; Oxelman, B.; Eriksson, J. S.; Pfeil, B. E. Phylogenetic Properties of 50 Nuclear Loci in Medicago (Leguminosae) Generated Using Multiplexed Sequence Capture and Next-generation Sequencing. *PLoS One* **2014,** *9* (10), e109704. https://doi.org/10.1371/journal.pone.0109704
Dereeper, A.; Nicolas, S.; Le Cunff, L.; Bacilieri, R.; Doligez, A.; Peros, J. P.; Ruiz, M.; This, P. SNiPlay: A Web-based Tool for Detection, Management and Analysis of SNPs. Application to Grapevine Diversity Projects. *BMC Bioinfo.* **2011,** *12*, 134. https://doi.org/10.1186/1471-2105-12-134
Dohm, J. C.; Lottaz, C.; Borodina, T.; Himmelbauer, H. Substantial Biases in Ultra-short Read Data Sets from High-throughput DNA Sequencing. *Nucl. Acids Res.* **2008,** *36* (16), e105. https://doi.org/10.1093/nar/gkn425
El-Gebali, S.; Mistry, J.; Bateman, A.; Eddy, S. R.; Luciani, A.; Potter, S. C.; Qureshi, M.; Richardson, L. J.; Salazar, G. A.; Smart, A.; Sonnhammer, E. L. L.; Hirsh, L.; Paladin, L.; Piovesan, D.; Tosatto, S. C. E.; Finn, R. D. The Pfam Protein Families Database in 2019. *Nucl. Acids Res.* **2019,** *47* (D1), D427–D432. https://doi.org/10.1093/nar/gky995
Elena, S. F.; Fraile, A.; Garcia-Arenal, F. Evolution and Emergence of Plant Viruses. *Adv. Virus Res.* **2014,** *88*, 161–191. https://doi.org/10.1016/B978-0-12-800098-4.00003-9
Faircloth, B. C. Msatcommander: Detection of Microsatellite Repeat Arrays and Automated, Locus-specific Primer Design. *Mol. Ecol. Resour.* **2008,** *8* (1), 92–94. https://doi.org/10.1111/j.1471-8286.2007.01884.x
Favre, P.; Bapaume, L.; Bossolini, E.; Delorenzi, M.; Falquet, L.; Reinhardt, D. A Novel Bioinformatics Pipeline to Discover Genes Related to Arbuscular Mycorrhizal Symbiosis Based on Their Evolutionary Conservation Pattern among Higher Plants. *BMC Plant Biol.* **2014,** *14* (1), 333. https://doi.org/10.1186/s12870-014-0333-0

Ferguson, J.; Wheeler, W.; Fu, Y.; Prokunina-Olsson, L.; Zhao, H.; Sampson, J. Statistical Tests for Detecting Associations with Groups of Genetic Variants: Generalization, Evaluation, and Implementation. *Eur. J. Hum. Genet.* **2013,** *21* (6), 680–686. https://doi.org/10.1038/ejhg.2012.220

Forouzan, E.; Maleki, M. S. M.; Karkhane, A. A.; Yakhchali, B. Evaluation of Nine Popular de novo Assemblers in Microbial Genome Assembly. *J. Microbiol. Methods* **2017,** *143,* 32–37. https://doi.org/https://doi.org/10.1016/j.mimet.2017.09.008

Froissart, R.; Roze, D.; Uzest, M.; Galibert, L.; Blanc, S.; Michalakis, Y. Recombination Every Day: Abundant Recombination in a Virus during a Single Multi-Cellular Host Infection. *PLoS Biol.* **2005,** *3* (3), e89. https://doi.org/10.1371/journal.pbio.0030089

Gharizadeh, B.; Ghaderi, M.; Nyr, N. P. Pyrosequencing Technology for Short DNA Sequencing and Whole Genome Sequencing. *Seibutsu Butsuri* **2007,** *47* (2), 129–132. https://doi.org/10.2142/biophys.47.129

Gilad, Y.; Pritchard, J. K.; Thornton, K. Characterizing Natural Variation Using Next-Generation Sequencing Technologies. *Trends Genet.* **2009,** *25* (10), 463–471. https://doi.org/10.1016/j.tig.2009.09.003

Goldstein, D. B.; Weale, M. E. Population Genomics: Linkage Disequilibrium Holds the Key. *Curr. Biol.* **2001,** *11* (14), R576–579. https://doi.org/10.1016/s0960-9822(01)00348-7

Goodwin, S.; McPherson, J. D.; McCombie, W. R. Coming of Age: Ten Years of Next-generation Sequencing Technologies. *Nat. Rev. Genet.* **2016,** *17* (6), 333–351. https://doi.org/10.1038/nrg.2016.49

Grover, J. W.; Bomhoff, M.; Davey, S.; Gregory, B. D.; Mosher, R. A.; Lyons, E. CoGe LoadExp+: A Web-based Suite That Integrates Next-generation Sequencing Data Analysis Workflows and Visualization. *Plant Direct* **2017,** *1* (2). https://doi.org/10.1002/pld3.8

Gupta, A. K.; Gupta, U. D. Next Generation Sequencing and Its Applications. *Animal Biotechnol.* **2014,** 345–367. https://doi.org/10.1016/b978-0-12-416002-6.00019-5

Hart, C.; Lipson, D.; Ozsolak, F.; Raz, T.; Steinmann, K.; Thompson, J.; Milos, P. M. Single-Molecule Sequencing. In *Single Molecule Tools: Fluorescence Based Approaches, Part A*, 2010; pp 407–430. https://doi.org/10.1016/s0076-6879(10)72002-4

Huber, C. D.; DeGiorgio, M.; Hellmann, I.; Nielsen, R. Detecting Recent Selective Sweeps While Controlling for Mutation Rate and Background Selection [https://doi.org/10.1111/mec.13351]. *Mol. Ecol.* **2016,** *25* (1), 142–156. https://doi.org/https://doi.org/10.1111/mec.13351

Hufford, K. M.; Krauss, S. L.; Veneklaas, E. J. Inbreeding and Outbreeding Depression in *Stylidium hispidum*: Implications for Mixing Seed Sources for Ecological Restoration. *Ecol. Evol.* **2012,** *2* (9), 2262–2273. https://doi.org/10.1002/ece3.302

Huse, S. M.; Huber, J. A.; Morrison, H. G.; Sogin, M. L.; Welch, D. M. Accuracy and Quality of Massively Parallel DNA Pyrosequencing. *Genome Biol.* **2007,** *8* (7), R143. https://doi.org/10.1186/gb-2007-8-7-r143

Hutchison, C. A., 3rd. DNA Sequencing: Bench to Bedside and Beyond. *Nucl. Acids Res.* **2007,** *35* (18), 6227–6237. https://doi.org/10.1093/nar/gkm688

Illumina. *History of Illumina Sequencing* 2020. https://sapac.illumina.com/science/technology/next-generation-sequencing/illumina-sequencing-history.html

Johnson, M. T.; Lajeunesse, M. J.; Agrawal, A. A. Additive and Interactive Effects of Plant Genotypic Diversity on Arthropod Communities and Plant Fitness. *Ecol. Lett.* **2006,** *9* (1), 24–34. https://doi.org/10.1111/j.1461-0248.2005.00833.x

Jueterbock, A.; Bostrom, C.; Coyer, J. A.; Olsen, J. L.; Kopp, M.; Dhanasiri, A. K. S.; Smolina, I.; Arnaud-Haond, S.; Van de Peer, Y.; Hoarau, G. The Seagrass Methylome Is Associated With Variation in Photosynthetic Performance among Clonal Shoots. *Front. Plant Sci.* **2020,** *11*, 571646. https://doi.org/10.3389/fpls.2020.571646

Kasianowicz, J. J.; Bezrukov, S. M. On 'Three Decades of Nanopore Sequencing'. *Nat. Biotechnol.* **2016,** *34* (5), 481–482. https://doi.org/10.1038/nbt.3570

Kessler, A.; Halitschke, R.; Baldwin, I. T. Silencing the Jasmonate Cascade: Induced Plant Defenses and Insect Populations. *Science* **2004,** *305* (5684), 665–668. https://doi.org/10.1126/science.1096931

Koboldt, D. C.; Chen, K.; Wylie, T.; Larson, D. E.; McLellan, M. D.; Mardis, E. R.; Weinstock, G. M.; Wilson, R. K.; Ding, L. VarScan: Variant Detection in Massively Parallel Sequencing of Individual and Pooled Samples. *Bioinformatics* **2009,** *25* (17), 2283–2285. https://doi.org/10.1093/bioinformatics/btp373

Kofler, R.; Schlotterer, C.; Lelley, T. SciRoKo: A New Tool for Whole Genome Microsatellite Search and Investigation. *Bioinformatics* **2007,** *23* (13), 1683–1685. https://doi.org/10.1093/bioinformatics/btm157

Kono, N.; Arakawa, K. Nanopore Sequencing: Review of Potential Applications in Functional Genomics. *Dev. Growth Differ.* **2019,** *61* (5), 316–326. https://doi.org/10.1111/dgd.12608

Kress, W. J.; Erickson, D. L. A Two-Locus Global DNA Barcode for Land Plants: The Coding rbcL Gene Complements the Non-coding trnH-psbA Spacer Region. *PLoS One* **2007,** *2* (6), e508. https://doi.org/10.1371/journal.pone.0000508

Kulski, J. K. Next-Generation Sequencing—An Overview of the History, Tools, and "Omic" Applications. In *Next Generation Sequencing—Advances, Applications and Challenges 2016*. https://doi.org/10.5772/61964

Kumar, K. R.; Cowley, M. J.; Davis, R. L. Next-Generation Sequencing and Emerging Technologies. *Semin Thromb Hemost* **2019,** *45* (7), 661–673. https://doi.org/10.1055/s-0039-1688446

Kutnjak, D.; Elena, S. F.; Ravnikar, M. Time-Sampled Population Sequencing Reveals the Interplay of Selection and Genetic Drift in Experimental Evolution of Potato Virus Y. *J. Virol.* **2017,** *91* (16). https://doi.org/10.1128/JVI.00690-17

Lee, H.; Gurtowski, J.; Yoo, S.; Nattestad, M.; Marcus, S.; Goodwin, S.; Richard McCombie, W.; Schatz, M. C. Third-generation Sequencing and the Future of Genomics. *bioRxiv* **2016,** 048603. https://doi.org/10.1101/048603

Liu, L.; Du, Y.; Shen, C.; Li, R.; Lee, J.; Li, P. The Complete Chloroplast Genome of *Papaver setigerum* and Comparative Analyses in Papaveraceae. *Genet. Mol. Biol.* **2020,** *43* (3), e20190272. https://doi.org/10.1590/1678-4685-GMB-2019-0272

Liu, L. X.; Du, Y. X.; Folk, R. A.; Wang, S. Y.; Soltis, D. E.; Shang, F. D.; Li, P. Plastome Evolution in Saxifragaceae and Multiple Plastid Capture Events Involving Heuchera and Tiarella. *Front. Plant Sci.* **2020,** *11*, 361. https://doi.org/10.3389/fpls.2020.00361

Loman, N. J.; Misra, R. V.; Dallman, T. J.; Constantinidou, C.; Gharbia, S. E.; Wain, J.; Pallen, M. J. Performance Comparison of Benchtop High-throughput Sequencing Platforms. *Nat. Biotechnol.* **2012,** *30* (5), 434–439. https://doi.org/10.1038/nbt.2198

Luo, L.; Kong, X.; Gao, Z.; Zheng, Y.; Yang, Y.; Li, X.; Yang, D.; Geng, Y.; Yang, Y. Comparative Transcriptome Analysis Reveals Ecological Adaption of Cold Tolerance in Northward Invasion of *Alternanthera philoxeroides*. *BMC Genom.* **2020,** *21* (1), 532. https://doi.org/10.1186/s12864-020-06941-z

Lyons, E.; Pedersen, B.; Kane, J.; Alam, M.; Ming, R.; Tang, H.; Wang, X.; Bowers, J.; Paterson, A.; Lisch, D.; Freeling, M. Finding and Comparing Syntenic Regions among Arabidopsis and the Outgroups Papaya, Poplar, and Grape: CoGe with Rosids. *Plant Physiol.* **2008,** *148* (4), 1772–1781. https://doi.org/10.1104/pp.108.124867

Madritch, M. D.; Hunter, M. D. Phenotypic Diversity Influences Ecosystem Functioning in an Oak Sandhills Community. *Ecology* **2002,** *83* (8), 2084–2090. https://doi.org/10.2307/3072039

Mardis, E. R. Next-Generation Sequencing Platforms. *Annu. Rev. Anal. Chem. (Palo Alto Calif)* **2013,** *6*, 287–303. https://doi.org/10.1146/annurev-anchem-062012-092628

Margulies et al., Genome Sequencing in Microfabricated High-density Picolitre Reactors. *Nature* **2005,** *437* (7057), 376–380. https://doi.org/10.1038/nature03959

Marth, G. T.; Korf, I.; Yandell, M. D.; Yeh, R. T.; Gu, Z.; Zakeri, H.; Stitziel, N. O.; Hillier, L.; Kwok, P. Y.; Gish, W. R. A General Approach to Single-nucleotide Polymorphism Discovery. *Nat. Genet.* **1999,** *23* (4), 452–456. https://doi.org/10.1038/70570

Massoumou, M.; van Tuinen, D.; Chatagnier, O.; Arnould, C.; Brechenmacher, L.; Sanchez, L.; Selim, S.; Gianinazzi, S.; Gianinazzi-Pearson, V. *Medicago truncatula* Gene Responses Specific to Arbuscular Mycorrhiza Interactions with Different Species and Genera of Glomeromycota. *Mycorrhiza* **2007,** *17* (3), 223–234. https://doi.org/10.1007/s00572-006-0099-9

Maxam, A. M.; Gilbert, W. A New Method for Sequencing DNA. *Proc. Natl. Acad. Sci. USA* **1977,** *74* (2), 560–564. https://doi.org/10.1073/pnas.74.2.560

McCarthy, E. W.; Berardi, A. E.; Smith, S. D.; Litt, A. Related Allopolyploids Display Distinct Floral Pigment Profiles and Transgressive Pigments. *Am. J. Bot.* **2017,** *104* (1), 92–101. https://doi.org/10.3732/ajb.1600350

McKain, M. R.; Johnson, M. G.; Uribe-Convers, S.; Eaton, D.; Yang, Y. Practical Considerations for Plant Phylogenomics. *Appl. Plant Sci.* **2018,** *6* (3), e1038. https://doi.org/10.1002/aps3.1038

McVean, G. The Structure of Linkage Disequilibrium Around a Selective Sweep. *Genetics* **2007,** *175* (3), 1395–1406. https://doi.org/10.1534/genetics.106.062828

Meglecz, E.; Costedoat, C.; Dubut, V.; Gilles, A.; Malausa, T.; Pech, N.; Martin, J. F. QDD: A User-friendly Program to Select Microsatellite Markers and Design Primers from Large Sequencing Projects. *Bioinformatics* **2010,** *26* (3), 403–404. https://doi.org/10.1093/bioinformatics/btp670

Merriman, B.; Ion Torrent, R.; Team, D.; Rothberg, J. M. Progress in Ion Torrent Semiconductor Chip Based Sequencing. *Electrophoresis* **2012,** *33* (23), 3397–3417. https://doi.org/10.1002/elps.201200424

Metzker, M. L. Sequencing Technologies—the Next Generation. *Nat. Rev. Genet.* **2010,** *11* (1), 31–46. https://doi.org/10.1038/nrg2626

Miller, M. R.; Dunham, J. P.; Amores, A.; Cresko, W. A.; Johnson, E. A. Rapid and Cost-effective Polymorphism Identification and Genotyping Using Restriction Site Associated DNA (RAD) Markers. *Genome Res.* **2007,** *17* (2), 240–248. https://doi.org/10.1101/gr.5681207

Mohammadi-Dehcheshmeh, M.; Niazi, A.; Ebrahimi, M.; Tahsili, M.; Nurollah, Z.; Ebrahimi Khaksefid, R.; Ebrahimi, M.; Ebrahimie, E. Unified Transcriptomic Signature of Arbuscular Mycorrhiza Colonization in Roots of *Medicago truncatula* by Integration of Machine Learning, Promoter Analysis, and Direct Merging Meta-Analysis [10.3389/fpls.2018.01550]. *Front. Plant Sci.* **2018,** *9*, 1550. https://www.frontiersin.org/article/10.3389/fpls.2018.01550

Morey, M.; Fernandez-Marmiesse, A.; Castineiras, D.; Fraga, J. M.; Couce, M. L.; Cocho, J. A. A Glimpse into Past, Present, and Future DNA Sequencing. *Mol. Genet. Metab.* **2013,** *110* (1–2), 3–24. https://doi.org/10.1016/j.ymgme.2013.04.024

Nelson, A. D. L.; Haug-Baltzell, A. K.; Davey, S.; Gregory, B. D.; Lyons, E. EPIC-CoGe: managing and analyzing genomic data. *Bioinformatics* **2018,** *34* (15), 2651–2653. https://doi.org/10.1093/bioinformatics/bty106

Nijveen, H.; van Kaauwen, M.; Esselink, D. G.; Hoegen, B.; Vosman, B. QualitySNPng: A User-friendly SNP Detection and Visualization Tool. *Nucl. Acids Res.* **2013,** *41* (Web Server issue), W587–W590. https://doi.org/10.1093/nar/gkt333

Novaes, E.; Drost, D. R.; Farmerie, W. G.; Pappas, G. J.; Jr.; Grattapaglia, D.; Sederoff, R. R.; Kirst, M. High-throughput Gene and SNP Discovery in *Eucalyptus grandis*, an Uncharacterized Genome. *BMC Genom.* **2008,** *9*, 312. https://doi.org/10.1186/1471-2164-9-312

Numa, H.; Itoh, T. MEGANTE: A Web-based System for Integrated Plant Genome Annotation. *Plant Cell Physiol.* **2014,** *55* (1), e2. https://doi.org/10.1093/pcp/pct157

Nyren, P.; Skarpnack, S. *Method of Sequencing DNA Based on the Detection of the Release of Pyrophosphate and Enzymatic Nucleotide Degradation* (US Patent No. USPTO), 2001. https://ip.com/pat/US6258568

Petri, K.; Fronza, R.; Gabriel, R.; Kappel, C.; Nowrouzi, A.; Linden, R. M.; Henckaerts, E.; Schmidt, M. Comparative Next-generation Sequencing of Adeno-associated Virus Inverted Terminal Repeats. *Biotechniques* **2014,** *56* (5), 269–273. https://doi.org/10.2144/000114170

Qi, J.; Zhao, F.; Buboltz, A.; Schuster, S. C. inGAP: An Integrated Next-generation Genome Analysis Pipeline. *Bioinformatics* **2010,** *26* (1), 127–129. https://doi.org/10.1093/bioinformatics/btp615

Rellstab, C.; Gugerli, F.; Eckert, A. J.; Hancock, A. M.; Holderegger, R. A Practical Guide to Environmental Association Analysis in Landscape Genomics. *Mol. Ecol.* **2015,** *24* (17), 4348–4370. https://doi.org/10.1111/mec.13322

Reuter, J. A.; Spacek, D. V.; Snyder, M. P. High-throughput Sequencing Technologies. *Mol. Cell* **2015,** *58* (4), 586–597. https://doi.org/10.1016/j.molcel.2015.05.004

Rhoads, A.; Au, K. F. PacBio Sequencing and Its Applications. *Genom. Proteom. Bioinfo.* **2015,** *13* (5), 278–289. https://doi.org/10.1016/j.gpb.2015.08.002

Ronaghi, M.; Karamohamed, S.; Pettersson, B.; Uhlen, M.; Nyren, P. Real-time DNA Sequencing Using Detection of Pyrophosphate Release. *Anal. Biochem.* **1996,** *242* (1), 84–89. https://doi.org/10.1006/abio.1996.0432

Ronaghi, M.; Nygren, M.; Lundeberg, J.; Nyren, P. Analyses of Secondary Structures in DNA by Pyrosequencing. *Anal. Biochem.* **1999,** *267* (1), 65–71. https://doi.org/10.1006/abio.1998.2978

Ross, M. G.; Russ, C.; Costello, M.; Hollinger, A.; Lennon, N. J.; Hegarty, R.; Nusbaum, C.; Jaffe, D. B. Characterizing and Measuring Bias in Sequence Data. *Genome Biol.* **2013,** *14* (5), R51. https://doi.org/10.1186/gb-2013-14-5-r51

Rothberg, J. M.; Hinz, W.; Rearick, T. M.; Schultz, J.; Mileski, W.; Davey, M.; Leamon, J. H.; Johnson, K.; Milgrew, M. J.; Edwards, M.; Hoon, J.; Simons, J. F.; Marran, D.; Myers, J. W.; Davidson, J. F.; Branting, A.; Nobile, J. R.; Puc, B. P.; Light, D.; Clark, T. A.; Huber, M.; Branciforte, J. T.; Stoner, I. B.; Cawley, S. E.; Lyons, M.; Fu, Y.; Homer, N.; Sedova, M.; Miao, X.; Reed, B.; Sabina, J.; Feierstein, E.; Schorn, M.; Alanjary, M.; Dimalanta, E.; Dressman, D.; Kasinskas, R.; Sokolsky, T.; Fidanza, J. A.; Namsaraev, E.; McKernan, K. J.; Williams, A.; Roth, G. T.; Bustillo, J. An Integrated Semiconductor Device Enabling Non-optical Genome Sequencing. *Nature* **2011,** *475* (7356), 348–352. https://doi.org/10.1038/nature10242

Roux, F.; Bergelson, J. The Genetics Underlying Natural Variation in the Biotic Interactions of Arabidopsis thaliana: The Challenges of Linking Evolutionary Genetics and Community Ecology. *Curr. Top. Dev. Biol.* **2016,** *119*, 111–156. https://doi.org/10.1016/bs.ctdb.2016.03.001

Sanger, F.; Air, G. M.; Barrell, B. G.; Brown, N. L.; Coulson, A. R.; Fiddes, C. A.; Hutchison, C. A.; Slocombe, P. M.; Smith, M. Nucleotide Sequence of Bacteriophage phi X174 DNA. *Nature* **1977,** *265* (5596), 687–695. https://doi.org/10.1038/265687a0

Sanger, F.; Coulson, A. R.; Friedmann, T.; Air, G. M.; Barrell, B. G.; Brown, N. L.; Fiddes, J. C.; Hutchison, C. A.; 3rd, Slocombe, P. M.; Smith, M. The Nucleotide Sequence of Bacteriophage phiX174. *J. Mol. Biol.* **1978,** *125* (2), 225–246. https://doi.org/10.1016/0022-2836(78)90346-7

Sanger, F.; Nicklen, S.; Coulson, A. R. DNA Sequencing with Chain-terminating Inhibitors. *Proc. Natl. Acad. Sci. USA* **1977,** *74* (12), 5463–5467. https://doi.org/10.1073/pnas.74.12.5463

Schweitzer, J. A.; Bailey, J. K.; Rehill, B. J.; Martinsen, G. D.; Hart, S. C.; Lindroth, R. L.; Keim, P.; Whitham, T. G. Genetically Based Trait in a Dominant Tree Affects Ecosystem Processes. *Ecol. Lett.* **2004,** *7* (2), 127–134. https://doi.org/https://doi.org/10.1111/j.1461-0248.2003.00562.x

Shendure, J.; Ji, H. Next-generation DNA sequencing. *Nat. Biotechnol. 2008, 26* (10), 1135–1145. https://doi.org/10.1038/nbt1486

Spigler, R. B. Plasticity of Floral Longevity and Floral Display in the Self-Compatible Biennial *Sabatia angularis* (Gentianaceae): Untangling the Role of Multiple Components of Pollination. *Ann. Bot.* **2017,** *119* (1), 167–176. https://doi.org/10.1093/aob/mcw195

Stockenhuber, R.; Zoller, S.; Shimizu-Inatsugi, R.; Gugerli, F.; Shimizu, K. K.; Widmer, A.; Fischer, M. C. Efficient Detection of Novel Nuclear Markers for Brassicaceae by Transcriptome Sequencing. *PLoS One* **2015,** *10* (6), e0128181. https://doi.org/10.1371/journal.pone.0128181

Thomas, D. Gene–Environment-Wide Association Studies: Emerging Approaches. *Nat. Rev. Genet.* **2010,** *11* (4), 259–272. https://doi.org/10.1038/nrg2764

Thomson, R. C.; Wang, I. J.; Johnson, J. R. Genome-enabled Development of DNA Markers for Ecology, Evolution and Conservation. *Mol. Ecol.* **2010,** *19* (11), 2184–2195. https://doi.org/10.1111/j.1365-294X.2010.04650.x

Treseder, K. K.; Vitousek, P. M. Potential Ecosystem-level Effects of Genetic Variation among Populations of *Metrosideros polymorpha* from a Soil Fertility Gradient in Hawaii. *Oecologia* **2001,** *126* (2), 266–275. https://doi.org/10.1007/s004420000523

Tromas, N.; Zwart, M. P.; Poulain, M.; Elena, S. F. Estimation of the In Vivo Recombination Rate for a Plant RNA Virus. *J. Gen. Virol.* **2014,** *95* (Pt 3), 724–732. https://doi.org/10.1099/vir.0.060822-0

Trucchi, E.; Mazzarella, A. B.; Gilfillan, G. D.; Lorenzo, M. T.; Schonswetter, P.; Paun, O. BsRADseq: Screening DNA Methylation in Natural Populations of Non-model Species. *Mol. Ecol.* **2016,** *25* (8), 1697–1713. https://doi.org/10.1111/mec.13550

Tyson, G. W.; Chapman, J.; Hugenholtz, P.; Allen, E. E.; Ram, R. J.; Richardson, P. M.; Solovyev, V. V.; Rubin, E. M.; Rokhsar, D. S.; Banfield, J. F. Community Structure and Metabolism through Reconstruction of Microbial Genomes from the Environment. *Nature* **2004,** *428* (6978), 37–43. https://doi.org/10.1038/nature02340

Ungerer, M. C.; Johnson, L. C.; Herman, M. A. Ecological Genomics: Understanding Gene and Genome Function in the Natural Environment. *Heredity (Edinb)* **2008,** *100* (2), 178–183. https://doi.org/10.1038/sj.hdy.6800992

van Dijk, E. L.; Jaszczyszyn, Y.; Naquin, D.; Thermes, C. The Third Revolution in Sequencing Technology. *Trends Genet.* **2018,** *34* (9), 666–681. https://doi.org/10.1016/j.tig.2018.05.008

Venter, J. C.; Remington, K.; Heidelberg, J. F.; Halpern, A. L.; Rusch, D.; Eisen, J. A.; Wu, D.; Paulsen, I.; Nelson, K. E.; Nelson, W.; Fouts, D. E.; Levy, S.; Knap, A. H.; Lomas, M. W.; Nealson, K.; White, O.; Peterson, J.; Hoffman, J.; Parsons, R.; Baden-Tillson, H.; Pfannkoch, C.; Rogers, Y. H.; Smith, H. O. Environmental Genome Shotgun Sequencing of the Sargasso Sea. *Science* **2004,** *304* (5667), 66–74. https://doi.org/10.1126/science.1093857

Wall, J. D.; Pritchard, J. K. Haplotype Blocks and Linkage Disequilibrium in the Human Genome. *Nat. Rev. Genet.* **2003,** *4* (8), 587–597. https://doi.org/10.1038/nrg1123

Wang, M. Z.; Li, H. L.; Li, J. M.; Yu, F. H. Correlations between Genetic, Epigenetic and Phenotypic Variation of an Introduced Clonal Herb. *Heredity (Edinb)* **2020,** *124* (1), 146–155. https://doi.org/10.1038/s41437-019-0261-8

Wang, P.; Moore, B. M.; Panchy, N. L.; Meng, F.; Lehti-Shiu, M. D.; Shiu, S. H. Factors Influencing Gene Family Size Variation Among Related Species in a Plant Family, Solanaceae. *Genome Biol. Evol.* **2018,** *10* (10), 2596–2613. https://doi.org/10.1093/gbe/evy193

Wang, Y.; Yang, Q.; Wang, Z. The Evolution of Nanopore Sequencing. *Front. Genet.* **2014,** *5*, 449. https://doi.org/10.3389/fgene.2014.00449

Weinig, C.; Ungerer, M. C.; Dorn, L. A.; Kane, N. C.; Toyonaga, Y.; Halldorsdottir, S. S.; Mackay, T. F.; Purugganan, M. D.; Schmitt, J. Novel Loci Control Variation in Reproductive Timing in *Arabidopsis thaliana* in Natural Environments. *Genetics* **2002,** *162* (4), 1875–1884. https://www.ncbi.nlm.nih.gov/pubmed/12524356

Whipple, A. V.; Holeski, L. M. Epigenetic Inheritance across the Landscape. *Front. Genet.* **2016,** *7*, 189. https://doi.org/10.3389/fgene.2016.00189

Wimp, G. M.; Martinsen, G. D.; Floate, K. D.; Bangert, R. K.; Whitham, T. G. Plant Genetic Determinants of Arthropod Community Structure and Diversity. *Evolution* **2005,** *59* (1), 61–69. https://www.ncbi.nlm.nih.gov/pubmed/15792227

Xiao, T.; Zhou, W. The Third Generation Sequencing: The Advanced Approach to Genetic Diseases. *Transl. Pediatr.* **2020,** *9* (2), 163–173. https://doi.org/10.21037/tp.2020.03.06

Yang, J.; Vazquez, L.; Chen, X.; Li, H.; Zhang, H.; Liu, Z.; Zhao, G. Development of Chloroplast and Nuclear DNA Markers for Chinese Oaks (Quercus Subgenus Quercus) and Assessment of Their Utility as DNA Barcodes. *Front. Plant Sci.* **2017,** *8*, 816. https://doi.org/10.3389/fpls.2017.00816

Yang, Z. PAML: A Program Package for Phylogenetic Analysis by Maximum Likelihood. *Comput. Appl. Biosci.* **1997,** *13* (5), 555–556. https://doi.org/10.1093/bioinformatics/13.5.555

Yang, Z. PAML 4: Phylogenetic Analysis by Maximum Likelihood. *Mol. Biol. Evol.* **2007,** *24* (8), 1586–1591. https://doi.org/10.1093/molbev/msm088

Zhang, J.; Wheeler, D. A.; Yakub, I.; Wei, S.; Sood, R.; Rowe, W.; Liu, P. P.; Gibbs, R. A.; Buetow, K. H. SNPdetector: A Software Tool for Sensitive and Accurate SNP Detection. *PLoS Comput. Biol.* **2005,** *1* (5), e53. https://doi.org/10.1371/journal.pcbi.0010053

Zhang, X.; Qin, H.; Xie, W.; Ma, Y.; Sun, W. Comparative Population Genetic Analyses Suggest Hybrid Origin of R Hododendron Pubicostatum, an Endangered Plant Species with Extremely Small Populations Endemic to Yunnan, China. *Plant Divers.* **2020,** *42* (4), 312–318. https://doi.org/10.1016/j.pld.2020.06.012

Zwart, M. P.; Elena, S. F. Matters of Size: Genetic Bottlenecks in Virus Infection and Their Potential Impact on Evolution. *Annu. Rev. Virol.* **2015,** *2* (1), 161–179. https://doi.org/10.1146/annurev-virology-100114-055135.

CHAPTER 4

Approaches for Metabarcoding and Environmental DNA (eDNA) Assays in Plants

MAHAK MAJEED and REIAZ UL REHMAN*

Department of Bioresources, University of Kashmir, Hazratbal, Srinagar, Jammu and Kashmir, India

Corresponding author. E-mail: rreiazbiores@gmail.com

ABSTRACT

The environmental DNA (eDNA) metabarcoding approach is an advanced analytical monitoring tool to access a diverse range of life forms on earth. eDNA derived from various biological samples in an environment (soil, sediments, and water) is analyzed for varied applications, such as determination of the abundance of species, population size, diet analysis, monitoring trade of invasive therapeutic floras, and in the conservation of threatened flora and fauna. In plant metabarcoding, rbcL, matK, and ITS have been assigned as standard markers that showed better resolution between closely related species due to their universality, better sequence, and discriminating qualities. Besides these markers, plastid genomes are also dominantly used in plant metabarcoding. Moreover, recent research has devised efficient database analysis programs, such as BOLD, GenGIS, QIIME, and MOTHUR that can overcome the problems of fewer accuracies arising during sequencing protocols. This chapter discusses the various potential attributes and recent advancements adopted in eDNA

Plant Ecogenomics. Peerzada Arshid Shabir, PhD, Peerzada Yasir Yousuf, PhD & Khalid Rehman Hakeem, PhD (Eds.)

metabarcoding approaches applied on various taxonomic groups most prevalently plants.

4.1 INTRODUCTION

The term DNA barcoding or metabarcoding is specifically applied to samples containing a multitude of organisms. Although there are uniform databases used as a reference in both barcoding and metabarcoding. The latter, in particular, utilizes high-throughput sequencing (HTS) methods to detect a particular taxon in the sample containing a mixture of taxa (Taberlet et al., 2012; Cristescu, 2014). Plants are prevalently identified using methods of barcoding and metabarcoding. eDNA metabarcoding has revolutionized the method to unravel biodiversity that can be used to identify individual species (Dougherty et al., 2016; Sansom and Sassoubre, 2017). eDNA metabarcoding is a robust tool to identify the biodiversity present in the environment on an enormous scale besides giving taxonomic information of a specific set of target taxa (Deiner et al., 2016; Stat et al., 2017). eDNA encompasses total genetic material contained in various specimens collected from the environment including air, fresh or marine water streams, soil, cellular or extracted DNA, and even entire living bodies (Ficetola et al., 2008; Barnes and Turner, 2016). The living remains are important sources to provide genetic information of particular individuals (Gillett et al., 2008; Aylward et al., 2018; Monge et al., 2020). DNA derived from these noninvasive samples can be further analyzed to describe the abundance of species (Lukacs and Burnham, 2005; Ruell et al., 2009; Mondol et al., 2009), effective population size, density, sex of individuals, and diet (Perez et al., 2006; Jane˘cka et al., 2008, 2011; Borthakur et al., 2011; Farrell et al., 2000). Besides DNA from environmental samples can provide information about mating systems (Gottelli et al., 2007), genetic diversity or inbreeding coefficients within a population (Russello et al., 2004; Smith and Wang, 2014), transfer of genes (Ernest et al., 2003; Haag et al., 2010), and evolutionarily significant units (Sharma et al., 2011). To ascertain the biodiversity at the level of species, eDNA is used as an efficient tool instead of traditional methods like trawl nets, trapping, and transects (Boussarie et al., 2018; Thomsen et al., 2016; Smart et al.,

2016; Port et al., 2016). At times of detection of low-density species, the eDNA methods are most durable for detecting the presence of low density, endangered, cryptic, and transient species (Rees et al., 2014; Ardura et al., 2015; Smith et al., 2006; Uthicke et al., 2018; Yamanaka and Minamoto, 2016; Gargan et al., 2017). As eDNA offers many advantages over traditional methodologies in terms of sensitivity and simplicity, it is used as a valuable biodiversity monitoring assessment tool especially for population genetics to reduce the errors of invasive sampling (McKelvey et al., 2016; Wilcox et al., 2016; Snyder, 2003; Kim et al., 2018). eDNA is an emerging analytical method utilized for the detection of eukaryotic micro-and microbial ecosystems (Ficetola et al., 2008; Thomsen et al., 2012; Zamor et al., 2012). However, the important utility of the method has been found in the detection of pathogens found in freshwater bodies (Gomes et al., 2017). eDNA metabarcoding is the comparatively novel, rapid, and economically viable biomonitoring method as it involves the simultaneous identification of a diverse spectrum of species from environmental samples (soil, water, sediments) with the aid of HTS (Thomsen and Willerslev, 2015) (Fig. 4.1).

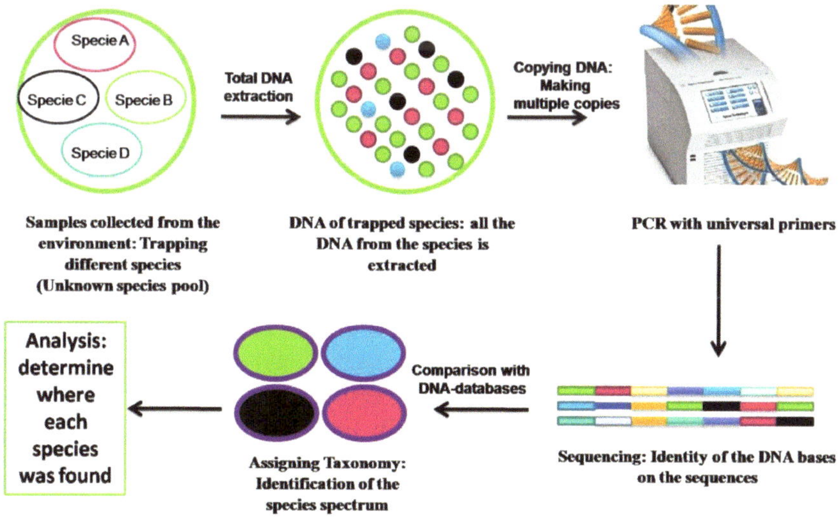

FIGURE 4.1 Metagenomics (eGenomics or community genomics) analysis of different species by DNA extraction for accessing it through genomic and bioinformatic tools for identification and analysis.

4.2 APPROACHES OF METABARCODING

The methodology employed in metabarcoding relies on the determination of specific taxa from the samples containing mixed organisms. Metabarcoding involves the selection of multiple loci, that are optimized for attaining the detection of different taxa at the species level. For instance, ITS is employed to identify fungi, matK/rbcL used to assess the diversity in plants, CO1 is employed to determine the genetic diversity in animals (Table 4.1). For the identification of individual specimens, the data generated from sequencing genes can be linked together in barcoding that helps in improving efficiency in the classification of taxa (Li et al., 2017). The large sample size used in metabarcoding prevents such treatment. However, in metabarcoding involving multiple loci, each result must be analyzed separately which is later combined to generate the group of taxa that share close resemblance from the total analyzed group (Arulandhu et al., 2017). Moreover, standardized multiple primer sets are needed due to variation in primer binding sites around metabarcoding loci for the amplification of various taxa. However, the mismatching of these primer binding sites in a sample can cause failure in the identification of desired species (Miya et al., 2015). The selection of locus is the determining factor of the quality of DNA recovered from the sample. Longer loci are chosen in the case of living samples like microbiota of soil which enables to obtain good quality DNA. The dead remnants of organisms represent the poor sources of DNA due to large amounts of degraded DNA which cause potential challenges in the amplification of those obtained DNA sequences. Thus, various alterations in standard barcoding loci target only those sequences of degraded DNA which are brief and significant from a taxonomic point of view (García-Robledo et al., 2013; de Vere et al., 2017; Erickson et al., 2017). Various factors are affecting the precision of quantification, few are linked to loci, such as primer mismatches and repeating sequences in the genome (Angly et al., 2014; Elbrecht and Leese, 2015; Piñol et al., 2015; Deiner et al., 2017), while other factors are independent of loci, such as polymerase sequence affinity (Nichols et al., 2018), survey methods (Deiner et al., 2017), and DNA isolation protocols (Brooks et al., 2015). In 2009, the Consortium for the Barcode of Life (CBOL) employed a standard locus for plants. The concerned body of researchers selected rbcL and matK as markers of choice besides plastid and DNA markers. Markers, such as rbcL and matK could easily distinguish about 70% of species

TABLE 4.1 Application of Metabarcoding Assays in Biodiversity Assessment Involving the Respective Primer Sets of Target Organisms.

Metabarcoding assay	Primer set	Target	Gene	Primer sequence	Amplification length	References
Fish (16S)	Fish16SF/D	Fish	16S rRNA	5′GACCCTATGGAGCTTTAGAC-3′	~200	Berry et al. (2017)
	16S2R			5′CGCTGTTATCCCTADRGTAATC-3′		Deagle et al. (2007)
Crustacean (16S)	Crust16S_F(short)	Crustacea	16S rRNA	5′GGGACGATAAGACCCTATA-3′	~170	Berry et al. (2017)
	Crust16S_R(short)			5′ATTACGCTGTTATCCCTAAAG-3′		
Eukaryotes (COI)	mlCOIin F	Metazoa	COI	5′GGWACWGGWTGAACWGTWTAY-CCYCC-3′	~313	Leray et al. (2013)
	jgHCO2198			5′TAIACYTCIGGRTGICCRAARAAYCA-3′		Geller et al. (2013)
Eukaryotes (18S)	18SF1	Universal	18SrRNA	5′GCCAGTAGTCATATGCTTGTCT-3′	~350	Pochon et al. (2013)
	18SR400			5′GCCTGCTGCCTTCCTT-3′		
Plants (trnL)	c	Plants	Chloroplast trnL (UAA)	5′CGAAATCGGTAGACGCTACG-3′	569	Taberlet et al. (1991)
	d			5′GGGGATAGAGGGACTTGAAC-3′	51	Taberlet et al. (2007)
	g			5′GGGCAATCCTGAGCCAA-3′	280	Baamrane et al. (2012)
Molluscs	OlaND5-F	Latipes	NDF 5	5′TCTTTACTATAATCCTGGCAGTCC-TTATC-3′	108 bp	Tsuji et al. (2018)
	OlaND5-R			5′CTGCTGCTAACTCTTTTTGTTGTTC-3′		
	OlaND5-Pr			5′AATCTAACTGCTCGCAAAGTCCC-ACGACT-3′		
Chordata (16SrRNA)	Osa16S-F	Sakaizumii (Japanese rice fish)	16SrRNA	5′ATCTTCAAGTAGAGGTGACAGACCA-3′	136 bp	Tsuji et al. (2018)
	Osa16S-R			5′AACTCTCTTGATTTCTAGTCATTTG-TGTC-3′		
	Osa16S-Pr			5′TGGATAGAAGTTCAGCCTC-3′		

due to their universality, affordability, and improved discriminatory and sequence quality (CBOL Plant Working Group, 2009). However, there is increased research focused on other loci, some of which are predominantly used in plant barcoding to enhance the resolution of particular taxon under investigation (Kress and Erickson, 2007; Fazekas et al., 2008; Vijayan and Tsou, 2010; Hollingsworth et al., 2011; Li et al., 2015).

In seed plants, ITS has been assigned as a pivotal genetic marker (China Plant Bol Group et al., 2011) while as psbA-trnH is highly used as an accessory locus (Kress and Erickson, 2012). cpDNA loci showed variability within species (Byrne and Hankinson, 2012). These markers have been recognized as significant tools in determining the relationship at lower taxonomic levels and in the differentiation of species showing close resemblance and species belonging to the same genus (Shepherd et al., 2013; Anderson et al., 2016). With the advancement in sequencing technologies, abundant regions of the genome were subjected to barcoding to enhance specimen identification (e.g., Tonti-Filippini et al., 2017). Due to the specific patterns of resolution and bulk availability in specimen tissues, the plastid genome is currently used as the most dominant barcoding and metabarcoding loci in seed plants (Hollingsworth et al., 2016).

4.3 METHODOLOGY OF METABARCODING

4.3.1 eDNA COLLECTION

eDNA can be obtained from environmental samples that include skin, saliva, secretions, eggs, sperms, fruit, pollen, urine, blood, fecal matter, leaves, roots, and decaying bodies of larger animals (Taberlet et al., 2012; Bohmann et al., 2014; Barnes and Turner, 2016). The samples of eDNA are preserved, isolated, and then PCR amplified. The amplicons are later categorized based on information derived from their sequences (Deiner et al., 2015). The information derived from sequences aids in the identification and classification of species. The production of eDNA relies on various factors, such as biomass, feeding habit of the organism, age, space occupancies, physiology, and life history (Barnes and Turner, 2016; Goldberg et al., 2016; Hering et al., 2018). The crude conventional surveying methods of determining species richness and abundance caused habitat disturbances and destruction which also could not detect the small and confined species, thereby, reducing the probability of estimating the communities. In such cases, eDNA enhances the taxonomic resolution by detecting the diverse species and sampling greater diversity (Deiner et al., 2017). Although eDNA can target rare species but cannot determine the qualitative factors of the population such as the proportion of sex and state of health (Goldberg et al., 2016; Deiner et al., 2017).

4.3.2 DNA AND RNA EXTRACTION

DNA extraction was carried using 2.0 mm ZR Bashing Bead lysis tubes that contain 1 mL lysis buffer. DNA and RNA were extracted simultaneously. Using about 1.5% agarose gel and nanophotometer, the quality of extracted DNA/RNA was estimated. The two sequential DNAases treatments were given to RNA extracts for removing DNA molecules from RNA samples (Langlet et al., 2013). The purity of RNA samples was further examined by carrying out 50-cycle PCR using DNA primers. When PCR analysis failed to yield the amplified products, this verified the complete removal of DNA sequences from RNA aliquots RNA aliquots. The purified RNA was then subjected to reverse transcription using reverse transcriptase (SuperScript1 III). The extracted DNA, RNA, and synthesized cDNA were further separated and kept under cold storage until used for further analysis. The PCR amplification was run in two steps for all extracted DNA (eDNA) and RNA (eRNA) to generate a library (Pramanik and Li, 2002).

4.3.3 DNA AMPLIFICATION

The cDNA was amplified using DNA barcode marker region, rbcL (CBOL Plant Working Group, 2009). The rbcL region was amplified followed by two rounds of PCR amplification and then unique 5 bp tags were attached to differentiate samples that can easily be attained after sequencing. First PCR amplification involved the usage of universal primers rbcLaf and rbcLr590 (de Vere et al., 2012) which were ligated with adapter tails, such as GACGATGAGTCCTGAGGTATGTCACCACAAACAGAGA-CTAAAGC and GACGATGAGTCCTGAGGTAGTCCACCGCGTAGA-CATTCAT respectively. PCR was carried using template DNA and primers. The synthesis process was conducted in a thermal cycler that mediated the denaturation for 2 min at 95°C and also permitting the 30 cycles for 2 min at 95°C. The process was further carried out at 50°C for 90 s and 72°C for 40 s. Finally, the elongation steps were carried out for 5 min at 72°C and lastly for 10 s at 30°C (de Vere et al., 2012). The amplified products were visualized on 1% agarose gel. The brightest medium and faint bands were diluted in the following dilution trends 1/2000, 1/1000, and 1/500, respectively. For second PCR amplification, the diluted amplification sequence was taken as a template sequence. Similar to the first, the PCR reaction was repeated with 15 cycles. After following this methodology, each sample was twice amplified and the resulting amplified sequences were further separated with Qiagen, PCR purification kit (Hawkins et al., 2015).

4.3.4 AMPLICON SEQUENCING

The length and quality of amplified sequences were examined with the aid of a 5 bp tag in different samples. The sequence lacking the 5 bp tag was removed as it could not be known for the primer sequence. After removal of Tag and primer sequences and the sequences having a read length of 250 bp were thrown away. Two BLAST databases were generated using the rbcL sequences that provided 98% sequence identity information for the native angiosperms of any region (de Vere et al., 2012). Another database was created by utilizing the sequence data that was obtained from GenBank of all extracted chloroplast DNA. Using Megablast, the DNA sequences from environmental samples were simultaneously scored against both databases. Upon matching the sequences with the top bit score specifically, the sequence was recognized to belong to that species. Similarly, when the top bit score was recorded similarly for variable species that belong to the same genus, the result signified the genus. However, the top bit score from multiple genera was belonging to the same family, the result was made at the family level. The sequences were also blasted from multiple families that were not given any level of taxonomic recognition and hence considered unknown. The sequence results for individual environmental samples were initially sorted manually to retain only those recorded species found in that particular geographical area. When the species under study were not recorded from a geographical locale, then its sequence was classified at the genus level. The analysis of results derived from sequence amplification containing sequencing errors and taxa recorded from 10 or lesser sequences were omitted from further analysis (Coissac et al., 2012). High-throughput DNA barcoding and metabarcoding approach target bulk volume of data which in some cases cause major problems in data analysis, such as fewer accuracies in HTS data, sequences referring to those species that lack sequence data information, statistical analysis of ecological fluctuations, nonuniform sequence data arising from pseudogenes and spatiotemporal investigation of biodiversity. For addressing these problems in data analysis, research and development have provided well-recognized database analysis programs, such as BOLD (Ratnasingham et al., 2009) and GenGIS (Parks et al., 2013). The efficient tools are also designed for microbial ecosystems, namely, QIIME (Caporaso et al., 2010) and MOTHUR (Schloss et al., 2009). Another efficient tool based on the 16S gene for taxonomic assignments includes a ribosomal database classifier (Cole et al., 2014).

4.3.5 LIBRARY PREPARATION

During ligation or PCR steps (Saitoh et al., 2016; Elbrecht and Leese, 2015), DNA molecules are attached with adapters to form libraries (Batovska et al., 2018; Rennstam et al., 2018) that further binds with a cell for sequencing. In metabarcoding studies, many samples are analyzed simultaneously on a single flow cell. It also incorporates the use of sequencing adapters to connect the sequencing oligonucleotides back to the original sample source. Although there are numerous indexing methods known for HTS (Gohl et al., 2016) but for diagnostic purposes, it is very indispensable to select an approach that can avoid unwanted recombination between molecules. The possible causes that can affect multiplexed sequencing include the formation of chimera during PCR (Schnell et al., 2015), excess quantities of primer/adapter oligonucleotide (Illumina, 2017), clusters formed on flow cell (Kircher et al., 2012), physical contamination during oligonucleotide synthesis or library preparation (Hanna and Doench, 2018; Nguyen et al., 2015). Index switching can cause taxa from one sample to extend to another leading to false-positive results for that particular taxon under investigation. When the probability of a true-positive result is ≥ 1, it implies that the particular taxon has greater geographical distribution than that existing in reality. Recent studies have demonstrated that instead of prominent combinatorial indexing, the application of unique dual indices is an effective method for regulating index switching. Application of unique indices leads to switching at both ends of the molecule generating an index combination. The mismatched indices to the sample are filtered which are excluded from analysis (Costello et al., 2018; Schnell et al., 2015; MacConaill et al., 2018). Indices sets are used alternatively for sequencing (Olmos et al., 2018). Therefore, to prevent sequence mismatches, index sequences applied should be designed with maximum edit distance between them such that substitution or insertion/deletion errors in index sequences cannot cause sequence mismatches (Wilcox et al., 2018; Faircloth and Glenn, 2012) especially in the case of higher error rate platforms such as nonopore (Krehenwinkel et al., 2019).

4.4 APPLICATIONS OF eDNA ASSAYS (BARCODING AND METABARCODING)

eDNA metabarcoding is a novel assessment approach that finds extensive applications in biodiversity estimates including the detection of invasive species (Comtet et al., 2015; Brown et al., 2016; Zhan et al., 2013; Pochon et al., 2013; Cristescu, 2014). After the death of an organism, DNA molecules showed prolonged existence in the environment which enables the detection of a diverse range of species such as those that are present in low densities and cannot be detected using conventional methods (Jerde et al., 2011). Metabarcoding has been also used in the quantification of airborne taxa in pollen or pollen bee loads (Kraaijeveld et al., 2015; Keller et al., 2015) at genus or family taxonomic levels. In the current scenario, metabarcoding has been widely applied to determine the taxonomic composition. It helps to evaluate the richness of plant or invertebrate communities (Epp et al., 2012; Fonseca et al., 2010; Ficetola et al., 2008). eDNA metabarcoding techniques have significantly impacted the evaluation of biomass and its abundance of the particular taxon (Deagle et al., 2013; Tillotson et al., 2018; Jungbluth et al., 2013; Diaz-Real et al., 2015; Lacoursière-Roussel et al., 2016). DNA barcodes attained from HTS as an important source of biological data about diverse taxa is applied in a new approach of ecosystem biomonitoring (Baird and Hajibabaei, 2012)

In the conservation and management of biodiversity, there has been great focus on generating DNA barcoding reference databases for plants such as the Terrestrial Ecosystem Research Network (TERN), with which it has become convenient to characterize ecosystems and ensuring their taxonomic integrity. These monitoring programs can be further augmented with many robust methodologies for biodiversity assessment (Thomsen and Willerslev, 2015; Bonin et al., 2018). The latest methods can be then validated using the data gathered from metabarcoding techniques (Aylagas et al., 2016). In conservation and management programs of biodiversity, identification of plant species remains a basic factor (Margules and Pressey, 2000), which involves the assessment of changing patterns of land use and development (Glasson and Therivel, 2013). In this regard, plant barcoding can provide a significant reference database that can be further used for molecular identification of plant species that are about to flower or produce fruits and in cases of less availability of plant tissue samples (Hollingsworth et al., 2016). Instead of morphological identification, molecular identification of plants is a rapid and cost-effective method to be used in biodiversity conservation and monitoring their environmental impact. Plant barcoding and metabarcoding also find important application in the assessment of trade of ornamental and agricultural floras (Dormontt et al., 2015; Ng et al., 2016; Staats et al., 2016). Barcoding represents a significant monitoring tool in the detection of the movement of invasive species and precious botanical products like native medicinal herbs (Newmaster et al., 2013; Seethapathy et al., 2015; Xin et al., 2015).

The other applications of metabarcoding approaches rely on elaborated database systems especially used to access a diversity of plants. These databases can be used to determine the gut contents of animals. This application also enables the identification of the threatened plant species from certain herbivorous animals (Nakahara et al., 2015; Nichols et al., 2016). The identified threatened species are then subjected to ex-situ and in-situ conservation measures after providing the food requirements of those threatened species (De Barba et al., 2014; Srivathsan et al., 2015). The application of metabarcoding on genetic material derived from the sediments of primitive times can help in revealing the old ecosystem links between species and ecosystem fluctuations (Thomsen and Willerslev, 2015; Parducci et al., 2017). Metabarcoding also holds significant implications in human health by evaluating the pollen loads and the allergen risks that can arise (Kraaijeveld et al., 2015).

4.5 CONCLUSION

In comparison to traditional sampling methods to evaluate biodiversity, eDNA surveys have received huge attention as an advanced, robust, and cost-effective biomonitoring tool. The significant potential of DNA metabarcoding lies in its capacity to investigate enormous amounts of the genome through the PCR process. As a result, this technique allows the simultaneous identification of genetic material obtained from numerous target species that act differently to PCR-based assays. DNA metabarcoding is a valuable technique in the identification of taxonomic links and the composition of diverse communities. This technique has been applied to detect the population size of even low abundant species to access the biodiversity on a magnified scale and formulating conservation strategies for endangered species. However, a lot of research still needs to be undertaken on designing group-specific metabarcoding primers and computer-based simulations for the development of the most appropriate primer pairs to extend their application on a broader range of taxonomic groups and ecoregions. Recent studies are also focused to enhance the applicability and efficiency of eDNA metabarcoding assays particularly for the amplification of less abundant species as compared with broad-scale metabarcoding that needs standardized eDNA isolation methods. Optimization of all these bottlenecks in eDNA assays targeting specific taxonomic groups can further address potential challenges of biodiversity assessment in large and inaccessible biomes.

KEYWORDS

- **markers**
- **monitoring**
- **biodiversity**
- **metabarcoding**
- **genomes**

REFERENCES

Anderson, B. M.; Barrett, M. D.; Krauss, S. L.; Thiele, K. Untangling a Species Complex of Arid Zone Grasses (Triodia) Reveals Patterns Congruent with Co-occurring Animals. *Mol. Phylogenet. Evol.* **2016,** *101*, 142–162.

Angly, F. E.; Dennis, P. G.; Skarshewski, A.; Vanwonterghem, I.; Hugenholtz, P.; Tyson, G. W. CopyRighter: A Rapid Tool for Improving the Accuracy of Microbial Community Profiles through Lineage-Specific Gene Copy Number Correction. *Microbiome* **2014,** *2* (1), 1–13.

Ardura, A.; Zaiko, A.; Martinez, J. L.; Samulioviene, A.; Semenova, A.; Garcia-Vazquez, E. eDNA and Specific Primers for Early Detection of Invasive Species–A Case Study on the Bivalve Rangiacuneata, Currently Spreading in Europe. *Marine Environ. Res.* **2015,** *112*, 48–55.

Arulandhu, A. J.; Staats, M.; Hagelaar, R.; Voorhuijzen, M. M.; Prins, T. W.; Scholtens, I.; BarretoCrespo, M. T. Development and Validation of a Multi-Locus DNA Metabarcoding Method to Identify Endangered Species in Complex Samples. *GigaScience* **2017,** *6* (10), gix080.

Aylagas, E.; Borja, Á.; Irigoien, X.; Rodríguez-Ezpeleta, N. Benchmarking DNA Metabarcoding for Biodiversity-based Monitoring and Assessment. *Front. Marine Sci.* **2016,** *3*, 96.

Aylward, M. L.; Sullivan, A. P.; Perry, G. H.; Johnson, S. E.; Louis Jr, E. E. An Environmental DNA Sampling Method for Aye-ayes from Their Feeding Traces. *Ecol. Evol.* **2018,** *8* (18), 9229–9240.

Baamrane, M. A. A.; Shehzad, W.; Ouhammou, A.; Abbad, A.; Naimi, M.; Coissac, E.; Znari, M. Assessment of the Food Habits of the Moroccan dorcas gazelle in M'Sabih Talaa, West Central Morocco, Using the trn L Approach. *PLoS One* **2012,** *7* (4), e35643.

Baird, D. J.; Hajibabaei, M. Biomonitoring 2.0: A New Paradigm in Ecosystem Assessment Made Possible by Next-generation DNA Sequencing. *Mol. Ecol.* **2012,** *21* (8), 2039–2044.

Barnes, M. A.; Turner, C. R. The Ecology of Environmental DNA and Implications for Conservation Genetics. *Conserv. Genet.* **2016,** *17* (1), 1–17.

Batovska, J.; Lynch, S. E.; Cogan, N. O. I.; Brown, K.; Darbro, J. M.; Kho, E. A.; Blacket, M. J. Effective Mosquito and Arbovirus Surveillance Using Metabarcoding. *Mol. Ecol. Resour.* **2018,** *18* (1), 32–40.

Berry, T. E.; Osterrieder, S. K.; Murray, D. C.; Coghlan, M. L.; Richardson, A. J.; Grealy, A. K.; Bunce, M. DNA Metabarcoding for Diet Analysis and Biodiversity: A Case Study Using the Endangered Australian Sea Lion (Neophoca cinerea). *Ecol. Evol.* **2017,** *7* (14), 5435–5453.

Bohmann, K.; Evans, A.; Gilbert, M. T. P.; Carvalho, G. R.; Creer, S.; Knapp, M.; De Bruyn, M. Environmental DNA for Wildlife Biology and Biodiversity Monitoring. *Trends Ecol. Evol.* **2014,** *29* (6), 358–367.

Bonin, A.; Taberlet, P.; Zinger, L.; Coissac, E. *Environmental DNA: For Biodiversity Research and Monitoring*; Oxford University Press: Oxford, UK, 2018.

Borthakur, U.; Barman, R. D.; Das, C.; Basumatary, A.; Talukdar, A.; Ahmed, M. F.; Bharali, R. Noninvasive Genetic Monitoring of Tiger (Panthera tigris tigris) Population of Orang National Park in the Brahmaputra Floodplain, Assam, India. *Eur. J. Wildlife Res.* **2011,** *57* (3), 603–613.

Boussarie, G.; Bakker, J.; Wangensteen, O. S.; Mariani, S.; Bonnin, L.; Juhel, J. B.;Vigliola, L. Environmental DNA Illuminates the Dark Diversity of Sharks. *Sci. Adv.* **2018,** *4* (5), eaap9661.

Brooks, J. P.; Edwards, D. J.; Harwich, M. D.; Rivera, M. C.; Fettweis, J. M.; Serrano, M. G.; Strauss, J. F. The Truth about Metagenomics: Quantifying and Counteracting Bias in 16S rRNA Studies. *BMC Microbiol.* **2015,** *15* (1), 1–14.

Brown, E. A.; Chain, F. J.; Zhan, A.; MacIsaac, H. J.; Cristescu, M. E. Early Detection of Aquatic Invaders Using Metabarcoding Reveals a High Number of Non-indigenous Species in Canadian Ports. *Diversity Distribut.* **2016,** *22* (10), 1045–1059.

Byrne, M.; Hankinson, M. Testing the Variability of Chloroplast Sequences for Plant Phylogeography. *Aust. J. Bot.* **2012,** *60* (7), 569–574.

Caporaso, J. G.; Kuczynski, J.; Stombaugh, J.; Bittinger, K.; Bushman, F. D.; Costello, E. K.; Huttley, G. A. QIIME Allows Analysis of High-throughput Community Sequencing Data. *Nat. Methods* **2010,** *7* (5), 335–336.

Coissac, E.; Riaz, T.; Puillandre, N. Bioinformatic Challenges for DNA Metabarcoding of Plants and Animals. *Mol. Ecol.* **2012,** *21* (8), 1834–1847.

Cole, J. R.; Wang, Q.; Fish, J. A.; Chai, B.; McGarrell, D. M.; Sun, Y.; Tiedje, J. M. Ribosomal Database Project: Data and Tools for High throughput rRNA Analysis. *Nucl. Acids Res.* **2014,** *42* (D1), D633–D642.

Comtet, T.; Sandionigi, A.; Viard, F.; Casiraghi, M. DNA (Meta) Barcoding of Biological Invasions: A Powerful Tool to Elucidate Invasion Processes and Help Managing Aliens. *Biol. Invasions* **2015,** *17* (3), 905–922.

Costello, M.; Fleharty, M.; Abreu, J.; Farjoun, Y.; Ferriera, S.; Holmes, L.; Vicente, G. Characterization and Remediation of Sample Index Swaps by Non-redundant Dual Indexing on Massively Parallel Sequencing Platforms. *BMC Genom.* **2018,** *19* (1), 1–10.

Cristescu, M. E. From Barcoding Single Individuals to Metabarcoding Biological Communities: Towards an Integrative Approach to the Study of Global Biodiversity. *Trends Ecol. Evol.* **2014,** *29* (10), 566–571. https://doi.org/10.1016/j.tree.2014.08.001

De Barba, M.; Miquel, C.; Boyer, F.; Mercier, C.; Rioux, D.; Coissac, E.; Taberlet, P. DNA Metabarcoding Multiplexing and Validation of Data Accuracy for Diet Assessment: Application to Omnivorous Diet. *Mol. Ecol. Resour.* **2014,** *14* (2), 306–323.

de Vere, N.; Jones, L. E.; Gilmore, T.; Moscrop, J.; Lowe, A.; Smith, D.; Ford, C. R. Using DNA Metabarcoding to Investigate Honey Bee Foraging Reveals Limited Flower Use Despite High Floral Availability. *Sci. Rep.* **2017,** *7*, 42838.

de Vere, N.; Rich, T. C.; Ford, C. R.; Trinder, S. A.; Long, C.; Moore, C. W.; Tatarinova, T. DNA Barcoding the Native Flowering Plants and Conifers of Wales. *PloS One* **2012,** *7* (6), e37945.

Deagle, B. E.; Gales, N. J.; Evans, K.; Jarman, S. N.; Robinson, S.; Trebilco, R.; Hindell, M. A. Studying Seabird Diet through Genetic Analysis of Faeces: A Case Study on Macaroni Penguins (*Eudyptes chrysolophus*). *PLoS One* **2007,** *2* (9), e831.

Deagle, B. E.; Thomas, A. C.; Shaffer, A. K.; Trites, A. W.; Jarman, S. N. Quantifying Sequence Proportions in a DNA-based Diet Study Using Ion Torrent Amplicon Sequencing: Which Counts Count? *Mol. Ecol. Resour.* **2013,** *13* (4), 620–633.

Deiner, K.; Bik, H. M.; Mächler, E.; Seymour, M.; Lacoursière-Roussel, A.; Altermatt, F.; Pfrender, M. E. Environmental DNA Metabarcoding: Transforming How We Survey Animal and Plant Communities. *Mol. Ecol.* **2017,** *26* (21), 5872–5895.

Deiner, K.; Fronhofer, E. A.; Mächler, E.; Walser, J. C.; Altermatt, F. Environmental DNA Reveals That Rivers Are Conveyer Belts of Biodiversity Information. *Nat. Commun.* **2016a,** *7*, 12544. https://doi.org/10.1038/ncomms12544

Deiner, K.; Fronhofer, E. A.; Mächler, E.; Walser, J. C.; Altermatt, F. Environmental DNA Reveals That Rivers Are Conveyer Belts of Biodiversity Information. *Nat. Commun.* **2016b,** *7* (1), 1–9.

Deiner, K.; Walser, J. C.; Mächler, E.; Altermatt, F. Choice of Capture and Extraction Methods Affect Detection of Freshwater Biodiversity from Environmental DNA. *Biol. Conserv.* **2015,** *183*, 53–63.

Diaz-Real, J.; Serrano, D.; Píriz, A.; Jovani, R. (2015). NGS Metabarcoding Proves Successful for Quantitative Assessment of Symbiont Abundance: The Case of Feather Mites on Birds. *Exp. Appl. Acarol.* **2015,** *67* (2), 209–218.

Dormontt, E. E.; Boner, M.; Braun, B.; Breulmann, G.; Degen, B.; Espinoza, E.; Lee, S. L. Forensic Timber Identification: It's Time to Integrate Disciplines to Combat Illegal Logging. *Biol. Conserv.* **2015,** *191*, 790–798.

Dougherty, M. M.; Larson, E. R.; Renshaw, M. A.; Gantz, C. A.; Egan, S. P.; Erickson, D. M.; Lodge, D. M. Environmental DNA (eDNA) Detects the Invasive Rusty Crayfish *Orconectes rusticus* at Low Abundances. *J. Appl. Ecol.* **2016,** *53* (3), 722–732. https://doi.org/10.1111/1365-2664.12621

Elbrecht, V.; Leese, F. Can DNA-based Ecosystem Assessments Quantify Species Abundance? Testing Primer Bias and Biomass—Sequence Relationships with an Innovative Metabarcoding Protocol. *PloS One* **2015,** *10* (7), e0130324.

Epp, L. S.; Boessenkool, S.; Bellemain, E. P.; Haile, J.; Esposito, A.; Riaz, T.; Stenøien, H. K. New Environmental Metabarcodes for Analysing Soil DNA: Potential for Studying Past and Present Ecosystems. *Mol. Ecol.* **2012,** *21* (8), 1821–1833.

Erickson, D. L.; Reed, E.; Ramachandran, P.; Bourg, N. A.; McShea, W. J.; Ottesen, A. Reconstructing a Herbivore's Diet Using a Novel rbc L DNA Mini-barcode for Plants. *AoB Plants* **2017,** *9* (3), plx015.

Ernest, H. B.; Boyce, W. M.; Bleich, V. C.; May, B.; Stiver, S. J.; Torres, S. G. Genetic Structure of Mountain Lion (*Puma concolor*) Populations in California. *Conserv. Genet.* **2003,** *4* (3), 353–366.

Faircloth, B. C.; Glenn, T. C. Not All Sequence Tags Are Created Equal: Designing and Validating Sequence Identification Tags Robust to Indels. *PloS One* **2012,** *7* (8), e42543.

Farrell, L. E.; Roman, J.;Sunquist, M. E. Dietary Separation of Sympatric Carnivores Identified by Molecular Analysis of Scats. *Mol. Ecol.* **2000,** *9* (10), 1583–1590.

Fazekas, A. J.; Burgess, K. S.; Kesanakurti, P. R.; Graham, S. W.; Newmaster, S. G.; Husband, B. C.; Barrett, S. C. Multiple Multilocus DNA Barcodes from the Plastid Genome Discriminate Plant Species Equally Well. *PloS One* **2008,** *3* (7), e2802.

Ficetola, G. F.; Miaud, C.; Pompanon, F.; Taberlet, P. Species Detection Using Environmental DNA from Water Samples. *Biol. Lett.* **2008,** *4* (4), 423–425.

Fonseca, V. G.; Carvalho, G. R.; Sung, W.; Johnson, H. F.; Power, D. M.; Neill, S. P.; Creer, S. Second-generation Environmental Sequencing Unmasks Marine Metazoan Biodiversity. *Nat. Commun.* **2010,** *1* (1), 1–8.

García-Robledo, C.; Erickson, D. L.; Staines, C. L.; Erwin, T. L.; Kress, W. J. Tropical Plant–Herbivore Networks: Reconstructing Species Interactions Using DNA Barcodes. *PLoS One* **2013,** *8* (1), e52967.

Gargan, L. M.; Morato, T.; Pham, C. K.; Finarelli, J. A.; Carlsson, J. E.; Carlsson, J. Development of a Sensitive Detection Method to Survey Pelagic Biodiversity Using eDNA and Quantitative PCR: A Case Study of Devil Ray at Seamounts. *Marine Biol.* **2017,** *164* (5), 112.

Geller, J.; Meyer, C.; Parker, M.; Hawk, H. Redesign of PCR Primers for Mitochondrial Cytochrome C Oxidase Subunit I for Marine Invertebrates and Application in All-taxa Biotic Surveys. *Mol. Ecol. Resour.* **2013,** *13* (5), 851–861.

Gillett, R. M.; White, B. N.; Rolland, R. M. Quantification and Genetic Profiling of DNA Isolated from Free-Floating Feces of the North Atlantic Right Whale (Eubalaenaglacialis). *Marine Mammal Sci.* **2008,** *24* (2), 341–355.

Glasson, J.; Therivel, R. *Introduction to Environmental Impact Assessment*; Routledge: London, 2013.

Gohl, D. M.; Vangay, P.; Garbe, J.; MacLean, A.; Hauge, A.; Becker, A.; Knights, D. Systematic Improvement of Amplicon Marker Gene Methods for Increased Accuracy in Microbiome Studies. *Nat. Biotechnol.* **2016,** *34* (9), 942–949.

Goldberg, C. S.; Turner, C. R.; Deiner, K.; Klymus, K. E.; Thomsen, P. F.; Murphy, M. A.; Laramie, M. B. Critical Considerations for the Application of Environmental DNA Methods to Detect Aquatic Species. *Methods Ecol. Evol.* **2016,** *7* (11), 1299–1307.

Gomes, G. B.; Hutson, K. S.; Domingos, J. A.; Chung, C.; Hayward, S.; Miller, T. L.; Jerry, D. R. Use of Environmental DNA (eDNA) and Water Quality Data to Predict Protozoan Parasites Outbreaks in Fish Farms. *Aquaculture* **2017,** *479*, 467–473.

Gottelli, D.; Wang, J.; Bashir, S.; Durant, S. M. Genetic Analysis Reveals Promiscuity among Female Cheetahs. *Proc. R. Soc. B: Biol. Sci.* **2007,** *274* (1621), 1993–2001.

Group, C. P. B.; Li, D. Z.; Gao, L. M.; Li, H. T.; Wang, H.; Ge, X. J.; Yang, J. B. Comparative Analysis of a Large Dataset Indicates That Internal Transcribed Spacer (ITS) Should Be Incorporated Into the Core Barcode for Seed Plants. *Proc. Natl. Acad. Sci.* **2011,** *108* (49), 19641–19646.

Group, C. P. W.; Hollingsworth, P. M.; Forrest, L. L.; Spouge, J. L.; Hajibabaei, M.; Ratnasingham, S.; Fazekas, A. J. A DNA Barcode for Land Plants. *Proc. Natl. Acad. Sci.* **2009,** *106* (31), 12794–12797.

Haag, T.; Santos, A. S.; Sana, D. A.; Morato, R. G.; Cullen Jr, L.; CrawshawJr, P. G.; Eizirik, E. The Effect of Habitat Fragmentation on the Genetic Structure of a Top Predator: Loss of Diversity and High Differentiation among Remnant Populations of Atlantic Forest Jaguars (Panthera onca). *Mol. Ecol.* **2010,** *19* (22), 4906–4921.

Hanna, R. E.; Doench, J. G. A Case of Mistaken Identity. *Nat. Biotechnol.* **2018,** *36* (9), 802–804.

Hawkins, J.; De Vere, N.; Griffith, A.; Ford, C. R.; Allainguillaume, J.; Hegarty, M. J.; Adams-Groom, B. Using DNA Metabarcoding to Identify the Floral Composition of Honey: A New Tool for Investigating Honey Bee Foraging Preferences. *PLoS One* **2015,** *10* (8), e0134735.

Hering, D.; Borja, A.; Jones, J. I.; Pont, D.; Boets, P.; Bouchez, A.; Leese, F. Implementation Options for DNA-based Identification into Ecological Status Assessment under the European Water Framework Directive. *Water Res.* **2018,** *138*, 192–205.

Hollingsworth, P. M.; Graham, S. W.; Little, D. P. Choosing and Using a Plant DNA Barcode. *PloS One* **2011,** *6* (5), e19254.

Hollingsworth, P. M.; Li, D. Z.; van der Bank, M.; Twyford, A. D. Telling Plant Species Apart with DNA: From Barcodes to Genomes. *Phil. Trans. R. Soc. B Biol. Sci.* **2016,** *371* (1702), 20150338.

Illumina, I. Effects of Index Misassignment on Multiplexing and Downstream Analysis, 2017. www.illumina.com

Janečka, J. E.; Jackson, R.; Yuquang, Z.; Diqiang, L.; Munkhtsog, B.; Buckley-Beason, V.; Murphy, W. J. Population Monitoring of Snow Leopards Using Noninvasive Collection of Scat Samples: A Pilot Study. *Animal Conserv.* **2008,** *11* (5), 401–411.

Janečka, J. E.; Munkhtsog, B.; Jackson, R. M.; Naranbaatar, G.; Mallon, D. P.; Murphy, W. J. Comparison of Noninvasive Genetic and Camera-trapping Techniques for Surveying Snow Leopards. *J. Mammal.* **2011,** *92* (4), 771–783.

Jerde, C. L.; Mahon, A. R.; Chadderton, W. L.; Lodge, D. M. "Sight-unseen" Detection of Rare Aquatic Species Using Environmental DNA. *Conserv. Lett.* **2011,** *4* (2), 150–157.

Jungbluth, M. J.; Goetze, E.; Lenz, P. H. Measuring Copepod Naupliar Abundance in a Subtropical Bay Using Quantitative PCR. *Marine Biol.* **2013,** *160* (12), 3125–3141.

Keller, A.; Danner, N.; Grimmer, G.; Ankenbrand, M. V. D.; Von Der Ohe, K.; Von Der Ohe, W.; Steffan-Dewenter, I. Evaluating Multiplexed Next-generation Sequencing as a Method in Palynology for Mixed Pollen Samples. *Plant Biol.* **2015,** *17* (2), 558–566.

Kim, P.; Kim, D.; Yoon, T. J.; Shin, S. Early Detection of Marine Invasive Species, Bugulaneritina (Bryozoa: Cheilostomatida), Using Species-Specific Primers and Environmental DNA Analysis in Korea. *Marine Environ. Res.* **2018,** *139*, 1–10.

Kircher, M.; Sawyer, S.; Meyer, M. Double Indexing Overcomes Inaccuracies in Multiplex Sequencing on the Illumina Platform. *Nucl. Acids Res.* **2012,** *40* (1), e3–e3.

Kraaijeveld, K.; De Weger, L. A.; Ventayol García, M.; Buermans, H.; Frank, J.; Hiemstra, P. S.; Den Dunnen, J. T. Efficient and Sensitive Identification and Quantification of Airborne Pollen Using Next-Generation DNA Sequencing. *Mol. Ecol. Resour.* **2015,** *15* (1), 8–16.

Krehenwinkel, H.; Pomerantz, A.; Henderson, J. B.; Kennedy, S. R.; Lim, J. Y.; Swamy, V.; Prost, S. Nanopore Sequencing of Long Ribosomal DNA Amplicons Enables Portable and Simple Biodiversity Assessments with High Phylogenetic Resolution Across Broad Taxonomic Scale. *GigaScience* **2019,** *8* (5), giz006.

Kress, W. J.; Erickson, D. L. A Two-locus Global DNA Barcode for Land Plants: The Coding rbcL Gene Complements the Non-coding trnH-psbA Spacer Region. *PLoS One* **2007,** *2* (6), e508.

Kress, W. J.; Erickson, D. L. DNA Barcodes: Methods and Protocols. In *DNA Barcodes*; Humana Press: Totowa, NJ, 2012; pp 3–8.

Lacoursière-Roussel, A.; Côté, G.; Leclerc, V.; Bernatchez, L. Quantifying Relative Fish Abundance with eDNA: A Promising Tool for Fisheries Management. *J. Appl. Ecol.* **2016,** *53* (4), 1148–1157.

Langlet, D.; Geslin, E.; Baal, C.; Metzger, E.; Lejzerowicz, F.; Riedel, B.; Jorissen, F. J. Foraminiferal Survival after Long-term In Situ Experimentally Induced Anoxia. *Biogeosciences* **2013,** *10* (11), 7463.

Leray, M.; Yang, J. Y.; Meyer, C. P.; Mills, S. C.; Agudelo, N.; Ranwez, V.; Machida, R. J. A New Versatile Primer Set Targeting a Short Fragment of the Mitochondrial COI

Region for Metabarcoding Metazoan Diversity: Application for Characterizing Coral Reef Fish Gut Contents. *Front. Zoology* **2013,** *10* (1), 34.

Li, Q.; Wu, J.; Wang, Y.; Lian, X.; Wu, F.; Zhou, L.; Zhu, S. The Phylogenetic Analysis of Dalbergia (Fabaceae: Papilionaceae) Based on Different DNA Barcodes. *Holzforschung* **2017,** *71* (12), 939–949.

Li, X.; Yang, Y.; Henry, R. J.; Rossetto, M.; Wang, Y.; Chen, S. Plant DNA Barcoding: From Gene to Genome. *Biol. Rev.* **2015,** *90* (1), 157–166.

Lukacs, P. M.; Burnham, K. P. Review of Capture–Recapture Methods Applicable to Noninvasive Genetic Sampling. *Mol. Ecol.* **2005,** *14* (13), 3909–3919.

MacConaill, L. E.; Burns, R. T.; Nag, A.; Coleman, H. A.; Slevin, M. K.; Giorda, K.; Ducar, M. D. Unique, Dual-indexed Sequencing Adapters with UMIs Effectively Eliminate Index Cross-Talk and Significantly Improve Sensitivity of Massively Parallel Sequencing. *BMC Genom.* **2018,** *19* (1), 1–10.

Margules, C. R.; Pressey, R. L. Systematic Planning for Biodiversity Conservation. *Nature* **2000,** *405*, 243–253.

McKelvey, K. S.; Young, M. K.; Knotek, W. L.; Carim, K. J.; Wilcox, T. M.; Padgett-Stewart, T. M.; Schwartz, M. K. Sampling Large Geographic Areas for Rare Species Using Environmental DNA: A Study of Bull Trout Salvelinus confluentus Occupancy in Western Montana. *J. Fish Biol.* **2016,** *88* (3), 1215–1222.

Miya, M.; Sato, Y.; Fukunaga, T.; Sado, T.; Poulsen, J. Y.; Sato, K.; Kondoh, M. MiFish, a Set of universal PCR Primers for Metabarcoding Environmental DNA from Fishes: Detection of More than 230 Subtropical Marine Species. *R. Soc. Open Sci.* **2015,** *2* (7), 150088.

Mondol, S.; Karanth, K. U.; Kumar, N. S.; Gopalaswamy, A. M.; Andheria, A.; Ramakrishnan, U. Evaluation of Non-Invasive Genetic Sampling Methods for Estimating Tiger Population Size. *Biol. Conserv.* **2009,** *142* (10), 2350–2360.

Monge, O.; Dumas, D.; Baus, I. Environmental DNA from Avian Residual Saliva in Fruits and Its Potential Uses in Population Genetics. *Conserv. Genet. Resour.* **2020,** *12* (1), 131–139.

Nakahara, F.; Ando, H.; Ito, H.; Murakami, A.; Morimoto, N.; Yamasaki, M.; Isagi, Y. The Applicability of DNA Barcoding for Dietary Analysis of Sika Deer. *DNA Barcodes* **2015,** *3* (1), 200–206.

Newmaster, S. G.; Emailauthor, G. M.; Hanandh, S.; Ramalingam, S.; Ragupathy, S. DNA Barcoding Detects Contamination and Substitution in North American Herbal Products. *BMC Med.* **2013,** *20***1311,** 222.

Ng, K. K. S.; Lee, S. L.; Tnah, L. H.; Nurul-Farhanah, Z.; Ng, C. H.; Lee, C. T.; Khoo, E. Forensic Timber Identification: A Case Study of a CITES Listed Species, *Gonystylus bancanus* (Thymelaeaceae). *Forens. Sci. Int. Genet.* **2016,** *23*, 197–209.

Nguyen, N. H.; Smith, D.; Peay, K.; Kennedy, P. Parsing Ecological Signal from Noise in Next Generation Amplicon Sequencing. *New Phytol.* **2015,** *205* (4), 1389–1393.

Nichols, R. V.; Vollmers, C.; Newsom, L. A.; Wang, Y.; Heintzman, P. D.; Leighton, M.; Shapiro, B. Minimizing Polymerase Biases in Metabarcoding. *Mol. Ecol. Resour.* **2018,** *18* (5), 927–939.

Olmos, A.; Boonham, N.; Candresse, T.; Gentit, P.; Giovani, B.; Kutnjak, D.; Nakhla, M. K. High-throughput Sequencing Technologies for Plant Pest Diagnosis: Challenges and Opportunities. *EPPO Bull.* **2018,** *48* (2), 219–224.

Parks, D. H.; Mankowski, T.; Zangooei, S.; Porter, M. S.; Armanini, D. G.; Baird, D. J.; Beiko, R. G. GenGIS 2: Geospatial Analysis of Traditional and Genetic Biodiversity, with New Gradient Algorithms and an Extensible Plugin Framework. *PloS One* **2013,** *8* (7), e69885.

Perez, I.; Geffen, E.; Mokady, O. Critically Endangered Arabian Leopards Panthera pardus nimr in Israel: Estimating Population Parameters Using Molecular Scatology. *Oryx* **2006,** *40* (3), 295–301.

Piñol, J.; Mir, G.; Gomez-Polo, P.; Agustí, N. Universal and Blocking Primer Mismatches Limit the Use of High-throughput DNA Sequencing for the Quantitative Metabarcoding of Arthropods. *Mol. Ecol. Resour.* **2015,** *15* (4), 819–830.

Pochon, X.; Bott, N. J.; Smith, K. F.; Wood, S. A. Evaluating Detection Limits of Next-generation Sequencing for the Surveillance and Monitoring of International Marine Pests. *PloS One* **2013,** *8* (9), e73935.

Port, J. A.; O'Donnell, J. L.; Romero-Maraccini, O. C.; Leary, P. R.; Litvin, S. Y.; Nickols, K. J.; Kelly, R. P. Assessing Vertebrate Biodiversity in a Kelp Forest Ecosystem Using Environmental DNA. *Mol. Ecol.* **2016,** *25* (2), 527–541.

Pramanik, S.; Li, H. Direct Detection of Insertion/Deletion Polymorphisms in an Autosomal Region by Analyzing High-density Markers in Individual Spermatozoa. *Am. J. Hum. Genet.* **2002,** *71* (6), 1342–1352.

Rees, H. C.; Bishop, K.; Middleditch, D. J.; Patmore, J. R.; Maddison, B. C.; Gough, K. C. The Application of eDNA for Monitoring of the Great Crested Newt in the UK. *Ecol. Evol.* **2014,** *4* (21), 4023–4032.

Rennstam Rubbmark, O.; Sint, D.; Horngacher, N.; Traugott, M. A Broadly Applicable COI Primer Pair and an Efficient Single-Tube Amplicon Library Preparation Protocol for Metabarcoding. *Ecol. Evol.* **2018,** *8* (24), 12335–12350.

Ruell, E. W.; Riley, S. P.; Douglas, M. R.; Pollinger, J. P.; Crooks, K. R. Estimating Bobcat Population Sizes and Densities in a Fragmented Urban Landscape Using Noninvasive Capture–Recapture Sampling. *J. Mammal.* **2009,** *90* (1), 129–135.

Russello, M. A.; Gladyshev, E.; Miquelle, D.;Caccone, A. Potential Genetic Consequences of a Recent Bottleneck in the Amur Tiger of. *Conserv. Genet.* **2004,** *5* (5), 707–713.

Saitoh, S.; Aoyama, H.; Fujii, S.; Sunagawa, H.; Nagahama, H.; Akutsu, M.; Nakamori, T. A Quantitative Protocol for DNA Metabarcoding of Springtails (Collembola). *Genome* **2016,** *59* (9), 705–723.

Sansom, B. J.; Sassoubre, L. M. Environmental DNA (eDNA) Shedding and Decay Rates to Model Freshwater Mussel eDNA Transport in a River. *Environ. Sci. Technol.* **2017,** *51* (24), 14244–14253. https://doi.org/10.1021/acs.est.7b05199

Sansom, B. J.; Sassoubre, L. M. Environmental DNA (eDNA) Shedding and Decay Rates to Model Freshwater Mussel eDNA Transport in a River. *Environ. Sci. Technol.* **2017,** *51* (24), 14244–14253.

Schloss, P. D.; Westcott, S. L.; Ryabin, T.; Hall, J. R.; Hartmann, M.; Hollister, E. B.; Sahl, J. W. Introducing Mothur: Open-source, Platform-independent, Community-supported Software for Describing and Comparing Microbial Communities. *Appl. Environ. Microbiol.* **2009,** *75* (23), 7537–7541.

Schnell, I. B.; Bohmann, K.; Gilbert, M. T. P. Tag Jumps Illuminated–Reducing Sequence-to-Sample Misidentifications in Metabarcoding Studies. *Mol. Ecol. Resour.* **2015,** *15* (6), 1289–1303.

Seethapathy, G. S.; Ganesh, D.; Kumar, J. U. S.; Senthilkumar, U.; Newmaster, S. G.; Ragupathy, S.; Ravikanth, G. Assessing Product Adulteration in Natural Health Products for Laxative Yielding Plants, Cassia, Senna, and Chamaecrista, in Southern India Using DNA Barcoding. *Int. J. Legal Med.* **2015,** *129* (4), 693–700.

Sharma, R.; Stuckas, H.; Bhaskar, R.; Khan, I.; Goyal, S. P.; Tiedemann, R. Genetically Distinct Population of Bengal Tiger (Panthera tigris tigris) in Terai Arc Landscape (TAL) of India. *Mammal. Biol.* **2011,** *76* (4), 484–490.

Shepherd, K. A.; Perkins, A.; Collins, J.; Byrne, M.; Thiele, K. R. Morphological and Molecular Evidence Supports the Recognition of a New Subspecies of the Critically Endangered Pityro diascabra (Lamiaceae). *Aust. Syst. Bot.* **2013,** *26* (1), 1–12.

Smart, A. S.; Weeks, A. R.; van Rooyen, A. R.; Moore, A.; McCarthy, M. A.; Tingley, R. Assessing the Cost-efficiency of Environmental DNA Sampling. *Methods Ecol. Evol.* **2016,** *7* (11), 1291–1298.

Smith, M. A.; Woodley, N. E.; Janzen, D. H.; Hallwachs, W.; Hebert, P. D. DNA Barcodes Reveal Cryptic Host-specificity within the Presumed Polyphagous Members of a Genus of Parasitoid Flies (Diptera: Tachinidae). *Proc. Natl. Acad. Sci.* **2006,** *103* (10), 3657–3662.

Smith, O.; Wang, J. When Can Noninvasive Samples Provide Sufficient Information in Conservation Genetics Studies? *Mol. Ecol. Resour.* **2014,** *14* (5), 1011–1023.

Snyder, D. E. Invited Overview: Conclusions from a Review of Electrofishing and Its Harmful Effects on Fish. *Rev. Fish Biol. Fisheries* **2003,** *13* (4), 445–453.

Srivathsan, A.; Sha, J. C.; Vogler, A. P.; Meier, R. Comparing the Effectiveness of Metagenomics and Metabarcoding for Diet Analysis of a Leaf-Feeding Monkey (Pygathrix nemaeus). *Mol. Ecol. Resour.* **2015,** *15* (2), 250–261.

Staats, M.; Arulandhu, A. J.; Gravendeel, B.; Holst-Jensen, A.; Scholtens, I.; Peelen, T.; Kok, E. Advances in DNA Metabarcoding for Food and Wildlife Forensic Species Identification. *Analyt. Bioanalyt. Chem.* **2016,** *408* (17), 4615–4630.

Stat, M.; Huggett, M. J.; Bernasconi, R.; DiBattista, J. D.; Berry, T. E.; Newman, S. J.; Bunce, M. Ecosystem Biomonitoring with eDNA: Metabarcoding across the Tree of Life in a Tropical Marine Environment. *Sci. Rep.* **2017,** *7* (1), 1–11.

Sujeevan, R.; Hebert, P. A. U. L. BOLD: The Barcode of Life Data System. *Mol. Ecol. Notes* **2007,** *7*, 355–364.

Taberlet, P.; Coissac, E.; Pompanon, F.; Brochmann, C.; Willerslev, E. Towards Next-generation Biodiversity Assessment Using DNA Metabarcoding. *Mol. Ecol.* **2012,** *21* (8), 2045–2050.

Taberlet, P.; Coissac, E.; Pompanon, F.; Gielly, L.; Miquel, C.; Valentini, A.; Willerslev, E. Power and Limitations of the Chloroplast trn L (UAA) Intron for Plant DNA Barcoding. *Nucl. Acids Res.* **2007,** *35* (3), e14–e14.

Taberlet, P.; Gielly, L.; Pautou, G.; Bouvet, J. Universal Primers for Amplification of Three Non-Coding Regions of Chloroplast DNA. *Plant Mol. Biol.* **1991,** *17*, 1105–1109.

Thomsen, P. F.; Kielgast, J. O. S.; Iversen, L. L.; Wiuf, C.; Rasmussen, M.; Gilbert, M. T. P.;Willerslev, E. Monitoring Endangered Freshwater Biodiversity Using Environmental DNA. *Mol. Ecol.* **2012,** *21* (11), 2565–2573.

Thomsen, P. F.; Møller, P. R.; Sigsgaard, E. E.; Knudsen, S. W.; Jørgensen, O. A.; Willerslev, E. Environmental DNA from Seawater Samples Correlate with Trawl Catches of Subarctic, Deepwater Fishes. *PloS One* **2016,** *11* (11), e0165252.

Thomsen, P. F.; Willerslev, E. Environmental DNA–An Emerging Tool in Conservation for Monitoring Past and Present Biodiversity. *Biol. Conserv.* **2015,** *183*, 4–18.

Tillotson, M. D.; Kelly, R. P.; Duda, J. J.; Hoy, M.; Kralj, J.; Quinn, T. P. Concentrations of Environmental DNA (eDNA) Reflect Spawning Salmon Abundance at Fine Spatial and Temporal Scales. *Biol. Conserv.* **2018,** *220*, 1–11.

Tonti-Filippini, J.; Nevill, P. G.; Dixon, K.; Small, I. What Can We Do with 1000 Plastid Genomes? *Plant J.* **2017,** *90* (4), 808–818.

Tsuji, S.; Iguchi, Y.; Shibata, N.; Teramura, I.; Kitagawa, T.; Yamanaka, H. Real-Time Multiplex PCR for Simultaneous Detection of Multiple Species from Environmental DNA: An Application on Two Japanese Medaka Species. *Sci. Rep.* **2018,** *8* (1), 1–8.

Uthicke, S.; Lamare, M.; Doyle, J. R. eDNA Detection of Corallivorousseastar (Acanthaster cf. solaris) Outbreaks on the Great Barrier Reef Using Digital Droplet PCR. *Coral Reefs* **2018,** *37* (4), 1229–1239.

Vijayan, K.;Tsou, C. H. DNA Barcoding in Plants: Taxonomy in a New Perspective. *Curr. Sci.* **2010,** 1530–1541.

Wilcox, T. M.; McKelvey, K. S.; Young, M. K.; Sepulveda, A. J.; Shepard, B. B.; Jane, S. F.; Schwartz, M. K. Understanding Environmental DNA Detection Probabilities: A Case Study Using a Stream-Dwelling Char Salvelinus fontinalis. *Biological Conservation* **2016,** *194*, 209–216.

Wilcox, T. M.; Zarn, K. E.; Piggott, M. P.; Young, M. K.; McKelvey, K. S.; Schwartz, M. K. Capture Enrichment of Aquatic Environmental DNA: A First Proof of Concept. *Mol. Ecol. Resour.* **2018,** *18* (6), 1392–1401.

Xin, T.; Li, X.; Yao, H.; Lin, Y.; Ma, X.; Cheng, R.; Chen, S. Survey of Commercial Rhodiola Products Revealed Species Diversity and Potential Safety Issues. *Sci. Rep.* **2015,** *5*, 8337.

Yamanaka, H.; Minamoto, T. The Use of Environmental DNA of Fishes as an Efficient Method of Determining Habitat Connectivity. *Ecol. Indicators* **2016,** *62*, 147–153.

Zamor, R. M.; Glenn, K. L.; Hambright, K. D. Incorporating Molecular Tools into Routine HAB Monitoring Programs: Using qPCR to Track Invasive Prymnesium. *Harmful Algae* **2012,** *15*, 1–7.

Zhan, A.; Hulák, M.; Sylvester, F.; Huang, X.; Adebayo, A. A.; Abbott, C. L.; MacIsaac, H. J. High Sensitivity of 454 Pyrosequencing for Detection of Rare Species in Aquatic Communities. *Methods Ecol. Evol.* **2013,** *4* (6), 558–565.

CHAPTER 5

Role of Epigenetics in Ecogenomics

SHAHJAHAN RASHID[1], SUMIAH WANI[1], FARHANA WANI[1], ADIL RASHID NATH[2], and AFLAQ HAMID[1*]

[1]*Division of Plant Pathology, Sher-e-Kashmir University of Agricultural Science and Technology, Shalimar Srinagar, Kashmir, India*

[2]*Division of Plant Biotechnology, Sher-e-Kashmir University of Agricultural Science and Technology, Shalimar Srinagar, Kashmir, India*

[*]*Corresponding author. E-mail: falak19@gmail.com*

ABSTRACT

Epigenetics has been defined as the study of heritable changes in the expression of genes and function that cannot be explained by differences in the sequence of DNA. Researchers have enough shreds of evidence that phenotypes are not only governed by genomic composition and environmental inputs but there is another component that complements it, that is, epigenetics. Epigenetic mechanism (e.g., modifications of chromatin, methylation of DNA) can cause variation in traits having ecological importance in isolation of variation in genetics. It is now evident that some of the environmentally induced epigenetic variations in natural populations are autonomous from genetic variations and can be heritable to the next generations. To understand the mechanism behind natural variation in ecologically relevant traits it is necessary to take into account epigenetics. Studying ecological epigenetics will make us understand the mechanism

Plant Ecogenomics. Peerzada Arshid Shabir, PhD, Peerzada Yasir Yousuf, PhD & Khalid Rehman Hakeem, PhD (Eds.)

of how organisms respond to their environment. In this chapter, we have discussed some of the aspects of epigenetics in an ecological context.

5.1 INTRODUCTION

Genes or environmental exposures are not generally enough to explain completely a process but instead, they are one part of a larger complex molecular process that, as a whole, results in the phenotype. A better understanding of genotypic reaction to the external environmental conditions will make us understand how phenotype gets translated from genotype. Despite having ample knowledge on the sequence of several species, researchers have found it difficult to understand the genotype translation into complex phenotypes that are adaptive in the natural environment (Martin et al., 2011). Ecological epigenetics is one such area that may potentially contribute to understanding the relationship between epigenetic modification and ecologically important phenotypic traits. Epigenetics provides the link between genes and phenotype. In absence of variation in the DNA sequence, epigenetic processes can change the phenotypic outcome of a gene and organismal function (Richards, 2006) and can result in alteration of morphological characters (Cubas et al., 1999; Kucharski et al., 2008). Recent experimental observations have revealed that nongenomic variation in phenotypic can be stably incorporated and transferred to the next generation (Scoville et al., 2011), and theoretical studies have found that inheritance of such variation can influence the speed and direction of evolution (Lachmann and Jablonka, 1996; Day and Bonduriansky, 2011). It is now evident that there must be ecological and evolutionary implications of these epigenetic variations. This has created interest in characterizing the variation in natural populations which is often environment dependent and heritable (West-Eberhard, 2003; Hallgrımsson and Hall, 2012).

5.2 EPIGENETICS

Conrad Waddington coined the term epigenetics in the early 1940s. Waddington uses the term "epigenetics" in a broader context, to all developmental processes that determine how the phenotype gets translated from genotype (Holliday, 2006). Later it was Robin Holliday who recognized the concept of epigenetics and defined it as the study of alteration of expression of the gene and the mitotic inheritance of patterns of expression of the gene (Holliday, 1994; Jablonka and Lamb, 2002).

Recently, epigenetics has been defined as the "study of heritable changes in the expression of gene and function that are unexplainable by differences in the sequence of DNA" (Richards 2006; Bird 2007). Such epigenetic modifications can turn on, bring down, or inhibit the function of particular genes:

I. Cytosine methylation of DNA
II. Alteration of DNA packing via; chemical modification (addition of methyl or acetyl group) of histone proteins and
III. Mediation of regulatory processes by sRNA (small RNA) molecules.

These processes are dependent on each other and control gene function in an interactive, complex fashion (Berger, 2007).

Among the different epigenetic mechanisms which alter gene expression, in ecological epigenetics most-common mechanism studied is methylation of cytosines in DNA (incorporation of a methyl group to CpG islands) (Schrey et al., 2013). Methylation of these sites often results in decreased activity of the associated gene and some methyltransferase enzymes catalyze this process. There are enough pieces of evidence that epigenetic changes can transfer to future generations as in toadflax, Arabidopsis thaliana, and mice (Richards 2006; Blewitt et al., 2006; Vaughn et al., 2007; Bock 2012). Methyltransferase enzyme affects the transgenerational inheritance of DNA methylation in plants, which replicates patterns of methylation during both meiosis and mitosis (Takeda and Paszkowski, 2006).

A recent epigenetic study shows that natural variation in epigenetic modifications can be autonomous from DNA variation. For example, in Arabidopsis thaliana, large and consistent ecotypic variation in methylation of DNA has been found that was not associated with genetic variation (Cervera et al., 2002; Vaughn et al., 2007). Studies of polymorphism of DNA methylation by Keyte et al., (2006) in 20 accessions of cotton showed a high degree of epigenetic modifications than genetically based estimates of variation.

The difference between epigenetic and genetic variation is that sometimes epigenetic variation induced by the environment is transferred to the next generations (Richards 2006; Jirtle and Skinner, 2007). For example, phenotypes of four generations of progeny were affected by altering methylation of DNA in flax by Fieldes and Amyot (1999). In mice, the mating behavior of F3 generation got changed when treated with toxins that change DNA methylation (Crews et al., 2007). Also, changes in DNA methylation are induced by dietary supplementation, which is inherited to future generations (Cropley et al., 2006).

In species evolution and the biology of various invasive species, epigenetic variations may play a vital role as shreds of evidence suggest that these complement in hybridization and polyploidization events (Salmon et al., 2005; Chen and Ni, 2006).

DNA methylation is currently focused on ecological epigenetics but gene expression can also get affected by other epigenetic modifications. Histone modification results in the remodeling of DNA and alters its accessibility for transcription and can also interact with methylation of DNA (Rapp and Wendel, 2005). Small interfering RNAs (siRNAs) or methylation of DNA regulates the function of transposable elements (TEs), DNA regions, and incorporates them into new sites that are mobile within the genome (Kejnovsky et al., 2012; Slotkin et al., 2012). TEs when inserted within coding regions can change the expression of genes and function, so it is very important to regulate these regions of the genome (Feschotte, 2008). In histone and DNA methylation pathways siRNAs are active. In plants and animals, the similarities between these pathways indicate that these epigenetic processes are evolutionarily conserved (Saze et al., 2012). Broadly speaking, the evolutionary potential of species may increase by epigenetic processes due to abiotic stress and environmental challenges, which could be of utmost importance in climate change globally.

5.3 ECOLOGICAL EPIGENETICS

To know the importance of variation due to epigenetics and its inheritance in the natural environment, it is important to put epigenetics in an ecological context and their causes and effects must be studied in natural populations. Heritable phenotypic variations that are not explained by the difference in the sequence of DNA could be explained if epigenetic processes are taken into account. Thereby, enhancing our understanding of the mechanism behind ecologically relevant traits for their natural variation.

In a natural population, the epigenetic mechanism may have its role in microevolution in two ways. First, when heritable variation gets translated to phenotype and results in differences in fitness between organisms, this may provide another route for natural selection to transfer variations like genetic variation (Fig. 5.1). Second, unlike genetic variation, ecological interactions may directly change the epigenetic variation (Richards, 2006) and thereby provides another pathway to speed up the evolutionary change (Fig. 5.1). Ecogenomics has proved to be an important tool to investigate the genetic basis of the ecologically and evolutionary important phenotypic variations between and within the populations (Ungerer et al., 2008). Genetic variations are not enough to explain all ecologically relevant phenomena. Understanding phenotypic variation in populations mediated by epigenetic processes is what ecological epigenetic aims at (Bossdorf et al., 2008). One of the major attributes that distinguish ecological epigenetics from ecological genomics is that ecological epigenetics is generally more responsive and sensitive to the natural environment (Fig. 5.1) (Bossdorf et al., 2008; Angers et al., 2010).

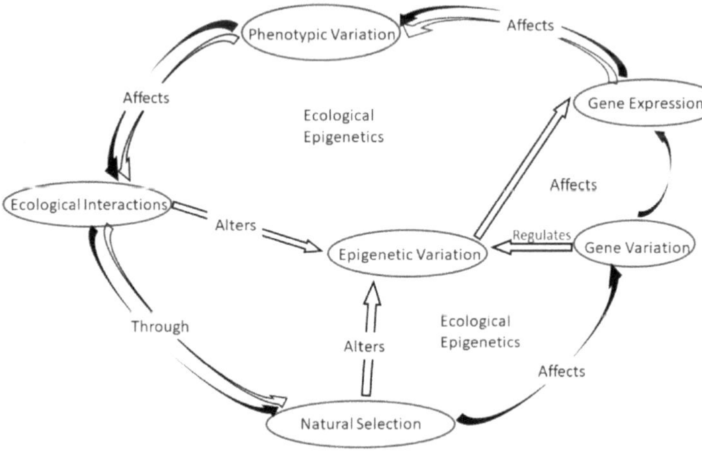

FIGURE 5.1 Interaction between ecological genetics (in solid arrows) and ecological epigenetics (in hollow arrows).

Since several developmental events have both genetic and epigenetic components, to some extent epigenetic variation is often accompanied by genetic variation among populations. Epigenetic variation can be entirely under genetic regulation in many of these cases, and thus its quantification would not offer much information. Epigenetic variation, however, can be partly or entirely independent from variation in the genome (Richards, 2006), and ecological epigenetics should concentrate on those situations.

5.4 EPIGENETIC VARIATION: EXTENT AND STRUCTURE WITHIN AND AMONG NATURAL POPULATIONS

In the external environment understanding the role of epigenetic processes is one of the most fundamental objectives underlying ecological epigenetics (Richards et al., 2012a). To procure this goal, the methylation patterns of DNA at individual and population levels are accessed by using different molecular approaches. Thus, amplified fragment length polymorphism (AFLP) markers that are sensitive to methylation (e.g., MseI is substituted by MspI and HpaII) are used in an ecological epigenetic context (Schrey et al., 2013). In many studies, this technique has been used to assess the degree and structure of population epigenetic variation. The MspI and HpaII enzymes have varied sensitivities to the CCGG recognition sequence of cytosine methylation (Roberts et al., 2007). The comparison of the banding pattern with MspI and HpaII from independent reactions enables the methylation state to be established at a specific restriction site (Salmon et al., 2008). At several restriction sites, methyl-sensitive amplified fragment length polymorphism (MS-AFLP) screens variations in methylation of DNA, producing a multilocus epigenotype for every organism. The use of MS-AFLP for studying variation in epigenetics in natural populations has many advantages (Schrey et al., 2013). First, this approach helps researchers to access epigenetic variation in organisms that have not yet been sequenced. This method is cost-effective and can be used in large sizes of samples for ecological studies. MS-AFLP also needs the same tools and technological expertise as conventional AFLP, making it possible for laboratories to link their genetic (i.e., AFLP) work with epigenetics. For population epigenetic studies, MS-AFLP is particularly useful as it offers a scan genome-wide and enables the screening of several organisms simultaneously at different loci. Finally, data of MS-AFLP procured from properly designed studies shows the heritable epigenetic variation, in particular, methylation of DNA can provide another variation source essential in natural selection.

5.5 INTERACTION BETWEEN EPIGENETIC VARIATION AND GENETIC VARIATION

Heritable trait variation is often contributed by epigenetic variations; however, if the variations in epigenetics are completely under the control of genetics then variation in epigenetics generally represents the underlying genetic changes (Richards 2006). As in A. thaliana where genetic regulation of variation in epigenetics has been demonstrated, in which changes in gene sequence associated with the mechanism of epigenetics may have a significant impact on the epigenotype (Becker et al., 2011). In model plant organisms, quantitative genetic studies also indicate that many variations in methylation of DNA between individuals are correlated with underlying differences in genetics. For instance, most differentially methylated regions (DMRs) were segregated stably in maize with the local allele DNA sequence (Li et al., 2014). In comparison, among 142 natural A. thaliana accessions, DMRs of only 35% were associated with polymorphisms of DNA sequence (Schmitz et al., 2013). Even though the relationship arising between local genetic polymorphisms and DMRs implicates the genetic control of methylation of DNA but without genetic control, such relationships can also occur during stable inheritance of spontaneous epimutation via epigenetic inheritance (Taudt et al., 2016).

Epigenetic variations, on the contrary, can affect also variation in genotype. Since methylation of DNA is correlated with the TEs silencing, through the regulation of TE function, epigenetic change may induce new genetic variation. DNA methylation reduction can lead to increased movement of TEs, causing variation in the sequence of DNA. This kind of dislocation of TEs influences phenotype through the induction of changes in the expression of a gene by attracting more methylation near it. DNA methylation of TEs in the RNA-mediated DNA methylation by well-characterized epialleles (Matzke and Mosher, 2014).

Various studies used less diverse non-model plant species for genetic variation control but they have not proved enough to suggest the relationship between epigenetic and genetic variation (Verhoeven et al., 2010; Preite and Verhoeven, 2014). Rather studies using statistical analysis to make conclusions of their independence by using similarities in DNA sequence as a factor to predict the corresponding similarities in methylation of DNA (Foust et al., 2016). Despite all of these observations, still they used molecular markers of low resolution, and thus the presence of critical genetic polymorphism was not ruled out (Dubin et al., 2015).

5.6 RATE OF SPONTANEOUS EPIMUTATIONS

Spontaneous epimutations contribute to heritable trait variations; however, the rate at which they occur and reverse determines their evolutionary potential. A study on accumulation of mutation in A. thaliana estimated epimutation rate, in which polymorphism of DNA methylation, accrued at a greater frequency than mutation in the genome (Ossowski et al., 2010; Van der Graaf et al., 2015). TEs have less occurrence of epimutation compared to the genic region. This pattern occurs frequently with the repair mechanism directed by RNA which inhibits the polymorphism development of methylation of DNA in TEs (Teixeira et al., 2009). At the level of DMR, spontaneous mutation also occurred which shows more functional relevance to the expression of the gene but compared to the rate of genetic mutation their frequency is low (Becker et al., 2011).

Till now, no data related to epimutation rates in the species of non-model plants have been found. A study by Verhoeven et al (2010) on apomictic dandelions revealed that not only in the variable environment but also in a steady environment, significant differences in methylation of DNA between individual plants developed, and most of it was transferred to future generations. Present pieces of evidence are not enough to ensure that the epimutations are stable and thus it requires wide and thorough study before such mutations can be given any significance related to their ecology and evolutionary tendencies.

5.7 EFFECT OF EPIGENETIC VARIATION ON PHENOTYPIC VARIATION IN ECOLOGICALLY IMPORTANT TRAITS

To study the effect of epigenetics on phenotype, model plants serve as a powerful tool to separate epigenetic effects on phenotype from genetic effects. In A. thaliana, epiRILs (epigenetic Recombinant Inbred Lines), derived from an initial cross of parents having the same genome that differ in methylation of DNA from each other are used to assess the phenotypic effects of variation in epigenetic in the complete absence of DNA sequence variation (Reinders et al., 2009). Such populations have largely enhanced our understanding involved in mechanisms of formation and stability of epialleles, in addition to the effects of alteration of methylation of DNA on phenotypic traits. The epiRIL populations showed significant phenotypic variation and DMRs were identified by linkage mapping in epiRILs that revealed heritable effects in phenotype for flowering time and root length (Cortijo et al., 2014). In natural populations of A. thaliana, a significant amount of the variable DMRs is also produced in the epiRILs, and in the wild, there is the possibility that they are also significant functionally.

In non-model species, experiments are required to confirm relationships between phenotypes and epigenetic variation. As some studies correlate among anonymous AFLP markers that are methylation-sensitive and leaf traits (Herrera and Bazaga, 2013), floral morphology (Herrera and Bazaga, 2010), and fitness-related traits (Medrano et al., 2014) in natural populations, but they could be simply the product of genetic variations.

5.8 ECOLOGICAL CONSEQUENCES OF EPIGENETIC VARIATION

Despite considerable progress to understand the degree of epigenetic modifications within and between populations, the extent to which epigenetic mechanism is linked to ecologically important traits has not been fully studied. There are some studies on ecology that correlated epigenotype to phenotype directly or that have evaluated how epigenetic effects mediates in phenotypic variations on organismal fitness. Epimutation of Linaria vulgaris is the classic example of epigenetic variations affecting ecological relevant traits (Cubas et al., 1999). In L. vulgaris, flower symmetry is regulated by the Lcycgene, epimutation causes silencing of this gene. Plants with hypomethylated epimutation result in radial symmetry flowers, while those with no epimutation present dorsoventral flower symmetry (Cubas et al., 1999). These epigenetic modifications in the morphology of flowers in L. vulgaris are heritable across generations and have significance in pollination and evolution.

As such we have not found any evidence that has indicated direct relation among epigenotype, functional trait, and stability in an ecological context. Various experimental investigations show that creating these relations could improve our insight into ecological processes. Given below are ecological important phenomena and how epigenetic processes may play a role in it.

5.8.1 RESPONSE TO ENVIRONMENTAL STIMULI

Being enough responsive to the environment is the only attribute that distinguishes epigenetics from genetic variation (Richards et al., 2010b). Lines of dandelion (Taraxacum officinale) that are genetically identical were used to evaluate the influence on epigenetic variation by five different biotic and abiotic conditions: salt stress, salicylic acid (to imitate damage of pathogen), low nutrients, jasmonic acid (to imitate damage of herbivore), and control treatment (Verhoeven et al., 2010). By usage of MS-AFLP, it was perused that change in methylation was large in treated plants genome wide than in control. Furthermore, it was also revealed that most of the modifications were inherited in the offsprings when grown in the common garden environment. To assess the changes on the epigenome by abiotic interactions Dowen et al. (2012) exposed A. thaliana to virulent bacterial strain, avirulent strain, and salicylic acid. Application of each treatment resulted in varied patterns of methylation, which demonstrate that stimuli of the environment are not only a healthy tool for global methylation but to every stimulus, the epigenome registered differently and regulates expression of gene dynamically (Dowen et al., 2012). Moreover, by differential methylation, TEs and the activity of their proximal genes are triggered, which supports the idea that dynamic TEs methylation may be a paramount element of the stress response.

5.8.2 NICHE BREADTH

To find out the mechanisms that manipulate the potential of some organisms for broader niche occupation within a community is also a milestone going in the field of ecology. Studies have revealed that the utilization of resources from a wide host range of environments especially when the environment is harsh is influenced by DNA methylation in Metschnikowia reukaufii (nectar living yeast) (Herrera et al., 2012). Herrera et al. (2012) applied demethylating agent 5-AzaC (5-Azacytidine) on yeast lines which are grown on media with different concentrations of sucrose, fructose, and glucose. The negative effect on growth by 5-AzaC was prominent in the harsh environmental conditions of high concentrations of sugar. The data from this experiment indicate that methylation of DNA in yeast reacts to varied conditions of nectar and the responses of epigenetic processes enable a diverse range of nectars and flowers to be occupied by the yeast.

5.8.3 INVASIVE SPECIES

Invasive species because of reduced genetic variation and the small size of the introduced population should encounter difficulty adapting to a new habitat. Despite facing the challenges, introduced species get adapted to new environments (Schrey et al., 2012). Several studies indicate that epigenetics may compensate for reduced genetic variation in the successful invasion of introduced species to new environments (Schrey et al., 2012). For instance, in Europe Fallopia japonica (Japanese knotweed) is an invasive plant species, now recently it has adapted to different habitats in the Northeastern United States. Samples taken from these diverse habitats have shown almost no genetic variation but a considerable amount of phenotypic plasticity, phenotypic and epigenetic variation, in response to controlled salt treatments (Richards et al., 2008, 2012b). These studies thus indicate that rapid colonization across a diverse environment in absence of genetic variation may be due to epigenetic variation. Thus, epigenetic modification may have potential in phenotypic diversification of introduced populations.

5.8.4 VARIATION IN BEHAVIOR

Among the various functional traits, behavior is generally regarded as highly responsive to the environment (West-Eberhard, 2003). Many studies have indicated that epigenetic processes mediate behavioral variation and phenotypic plasticity (Richard, 2011). For instance, in honey bee's epigenetic mechanism may affect the behavioral variation due to differential nutrition. Larvae feeding on a high-quality diet (royal jelly) develop into fertile queens which behave aggressively. While as, larvae taking lower quality diet to become infertile workers which spend most of their time foraging despite the same genome. Reduced DNA methylation in the brain of the queen was found when the DNA methylation levels were compared between queen and worker (Miklos and Maleszka, 2011). Furthermore, injecting larvae with siRNA results in suppression of methyltransferase system of DNA, which gave exact results as royal jelly diet (Miklos and Maleszka, 2011).

Another example of variation in behavior is from maternal care variation of rats which causes varied behavioral changes and stress reactivity that remains up to maturity (Weaver et al., 2004). Rats which receive little care from mother become fearful than rats which receive sufficient care. This is because the mother's care methylated differentially the glucocorticoid receptor (GR) promoter of the hippocampus of rats (Weaver et al., 2004). Cross-fostering shows that it is foster mother's behavior that determines the phenotype in young ones instead of the biological mother, which indicates epigenetic mechanism plays role in these, instead of genetic processes (Weaver et al., 2004).

5.8.5 DISEASE SUSCEPTIBILITY

An increasing body of evidence suggests that epigenetic processes responsive to the environment have been found to play a great role in regulating disease susceptibility (Tost, 2010). In one such study animals with the Avyallele, different methylation levels of TEs positioned upstream of the agouti gene have been found to play a role in expression. Hypo methylation results in yellow coat color and hypermethylation causes the brown color of wild type (Skinner et al., 2010). Moreover, obesity, diabetes, and tumor formation are associated with an unmethylated state (Skinner et al., 2010). Dietary supplementation of a mother with methyl donors and the genistein, phytoestrogen results in coat color change from yellow to brown in pseudoagouti offsprings and considerably decreases the incidence of cancer, diabetes, and obesity (Dolinoy et al., 2006, 2007). Thus, the effect of a mother's nutrition was directly linked to the vulnerability of disease in offspring and DNA methylation changes in the epigenome. This observation could have major ecological consequences for natural populations. Understanding the mechanism behind the response to nutrition and its effect on the disease will enhance our knowledge of the dynamics of disease.

Epigenomic variations arising from exposure to some chemicals (e.g., pesticides, fungicides) in the initial stages of life have been shown to influence the susceptibility of disease (Jirtle and Skinner 2007; Tost, 2010). For instance, different chemicals have been found to affect the various developmental stages of an organism like DNA methylation effect from bisphenol A (BPA) induces cancer susceptibility in rats (Skinner et al., 2010), while as fungicides and pesticides exposure results in reduced spermatogenic capacity and male infertility which were transmissible (Skinner et al., 2010). Even though the available data have enhanced our understanding of epigenetic processes in the environmental basis of disease, but there are no shreds of evidence that suggests as to what extent these processes influence disease susceptibility in the natural population.

5.9 EVOLUTIONARY CONSEQUENCES OF EPIGENETIC VARIATION

Various experimental findings have shown the importance of spontaneous epigenetic variation induced by the environment in evolutionary theory (Kronholm and Collins, 2016). Epigenetics has diversified the study of heritable phenotypes that occur in isolation of variation in DNA sequence. The evolutionary effect of epigenetic modification will rely on stability and the rate of epigenetic variations and distribution of their phenotypic effects (Kronholm and Collins, 2016). Modeling studies have revealed that if the rate of spontaneous epigenetic modifications is substantially higher than genetic variation, this could result in evolutionary dynamics, where epigenetic modifications cause a change in phenotype and ultimately become genetically encoded (Kronholm and Collins, 2016). These findings also revealed that epigenetic modifications that are transferred to the next generations could be adaptive in changing environments (Robertson and Richards 2015). The number of theoretical models proposed has shown the potential of epigenetic modification in changing the evolutionary dynamics. However, it is not sufficient we require further information to understand the role of epigenetics. For example, a study on A. thaliana has shown that there is not enough evidence for the persistence of epigenetic modifications across generations induced by the environment (Hagmann et al., 2015). To understand the role of epigenetics in evolutionary processes, it is important to have enough data related to how under diverse environmental conditions different species are affected epigenetically.

KEYWORDS

- **epigenetics**
- **DNA methylation**
- **ecological epigenetics**
- **MS-AFLP**
- **inheritance**
- **genetic variation**
- **phenotype**

REFERENCES

Angers, B.; Castonguay, E.; Massicotte, R. Environmentally Induced Phenotypes and DNA Methylation. How to Deal with Unpredictable Conditions Until the Next Generation and After. *Mol. Ecol.* **2010,** *19*(7), 1283–1295.

Becker, C.; Hagmann, J.; Muller, J.; Koenig, D.; Stegle, O.; Borgwardt, K.; et al. Spontaneous Epigenetic Variation in the Arabidopsis thaliana methylome. *Nature* **2011,** *480*, 245–249.

Berger, S. L. The Complex Language of Chromatin Regulation During Transcription. *Nature* **2007,** *447*, 407–412.

Bird, A. Perceptions of Epigenetics. *Nature* **2007,** *447*, 396–398.

Blewitt, M. E.; Vickaryous, N. K.; Paldi, A.; Koseki, H.; Whitelaw, E. Dynamic Reprogramming of DNA Methylation at an Epigenetically Sensitive Allele in Mice. *PLoS Genet.* **2006,** *2*, e49.

Bock, C. Analysing and Interpreting DNA Methylation Data. *Nat. Rev. Genet.* **2012,** *13*(10), 705–719.

Bossdorf, O.; Richards, C. L.; Pigliucci, M. Epigenetics for Ecologists. *Ecol. Lett.* **2008,** 11, 106–115

Cervera, M. T.; Ruiz-Garcia, L.; Martinez-Zapater, J. Analysis of DNA Methylation in Arabidopsis Thaliana Based on Methylation-Sensitive AFLP Markers. *Mol. Genet. Genomics* **2002**, *268*, 543–552.

Chen, Z. J.; Ni, Z. Mechanisms of Genomic Rearrangements and Gene Expression Changes in Plant Polyploids. *Bioessays* **2006**, *28*, 240–252.

Cortijo, S.; Wardenaar, R.; Colom_e-Tatch_e, M.; Gilly, A.; Etcheverry, M.; Labadie, K.; et al. Mapping the Epigenetic Basis of Complex Traits. *Science* **2014,** *343*(6175), 1145–1148.

Crews, D.; Gore, A. C.; Hsu, T. S.; Dangleben, N. L.; Spinetta, M.; Schallert, T.; et al. Transgenerational Epigenetic Imprints on Mate Preference. *Proc. Natl. Acad. Sci. USA* **2007**, *104*, 5942–5946.

Cropley, J. E.; Suter, C. M.; Beckman, K. B.; Martin, D. I. K. Germ-Line Epigenetic Modification of the Murine A(vy) Allele by Nutritional Supplementation. *Proc. Natl. Acad. Sci.* USA **2006,** *103*, 17308–17312.

Cubas, P.; Vincent, C.; Coen, E. An Epigenetic Mutation Responsible for Natural Variation in Floral Symmetry. *Nature* **1999,** *401*, 157–161.

Day, T.; Bonduriansky, R. A Unified Approach to the Evolutionary Consequences of Genetic and Nongenetic Inheritance. *Am. Nat.* **2011,** *178*, E18–E36.

Dolinoy, D. C.; Weidman, J. R.; Jirtle, R. L. Epigenetic Gene Regulation: Linking Early Developmental Environment to Adult Disease. *Reprod. Toxicol.* **2007,** *23*(3), 297–307.

Dolinoy, D. C.; Weidman, J. R.; Waterland, R. A.; Jirtle, R. L. Maternal Genistein Alters Coat Color and Protects Avy Mouse Offspring from Obesity by Modifying the Fetal Epigenome. *Environ. Heal Perspect.* **2006,** *144*(4), 567–572.

Dowen, R. H.; Pelizzola, M.; Schmitz, R. J.; Lister, R.; Dowen, J. M.; Nery, J. R.; Dixon, J. E.; Ecker, J. R. Widespread Dynamic DNA Methylation in Response to Biotic Stress. *Proc. Natl. Acad. Sci. USA* **2012,** *109*(32), E2183–E2191.

Dubin, M. J.; Zhang, P.; Meng, D.; Remigereau, M. S.; Osborne, E. J.; Casale, F. P.; et al. DNA Methylation Variation in Arabidopsis has a Genetic Basis and Shows Evidence of Local Adaptation. *eLife* **2015,** *4*, e05255. DOI: 10.7554/eLife.05255.

Feschotte, C. Transposable Elements and the Evolution of Regulatory Networks. *Nat. Rev. Genet.* **2008**, *9*, 397–405.

Fieldes, M. A.; Amyot, L. M. Epigenetic Control of Early Flowering in Flax Lines Induced by 5-Azacytidine Applied to Germinating Seed. *J. Hered.* **1999,** *90*, 199–206.

Foust, C. M.; Preite, V.; Schrey, A. W.; Verhoeven, K. J. F.; Richards, C. L. Genetic and Epigenetic Differences Associated with Environmental Gradients in Replicate Populations of Two Salt Marsh Perennials. *Mol. Ecol.* **2016,** *25*, 1639–1652.

Hagmann, J.; Becker, C.; Muller, J.; Stegle, O.; Meyer, R. C.; Wang, G.; et al. Century-Scale Methylome Stability in a Recently Diverged Arabidopsis thaliana Lineage. *PLoS Genet.* **2015,** *11*(1), e1004920.

Hallgrımsson, B.; Hall, B. K., Eds. Epigenetics: Linking Genotype and Phenotype in Development and Evolution; University of California Press: Berkeley (CA), 2012.

Herrera, C. M.; Bazaga, P. Epigenetic Differentiation and Relationship to Adaptive Genetic Divergence in Discrete Populations of the Violet Viola cazorlensis. *New Phytol.* **2010,** *187*(3), 867–876.

Herrera, C. M.; Bazaga, P. Epigenetic Correlates of Plant Phenotypic Plasticity: DNA Methylation Differs Between Prickly and Nonprickly Leaves in Heterophyllous Ilex aquifolium (Aquifoliaceae) Trees. *Bot. J. Linean Soc.* **2013,** *171*, 441–452.

Herrera, C. M.; Pozo, M. I.; Bazaga, P. Jack of all Nectars, Master of Most: DNA Methylation and the Epigenetic Basis of Niche Width in a Flower Living Yeast. *Mol. Ecol.* **2012,** *21*(11), 2602–2616.

Holliday, R. Epigenetics: An Overview. *Dev. Genet.* **1994,** *15*(6), 453–457.

Holliday, R. Epigenetics: A Historical Overview. *Epigenetics* **2006,** *1*(2), 76–80.

Jablonka, E.; Lamb, M. J. The Changing Concept of Epigenetics. *Ann N Y Acad. Sci.* **2002,** *981*, 82–96.

Jirtle, R. L.; Skinner, M. K. Environmental Epigenomics and Disease Susceptibility. *Nat. Rev. Genet.* **2007,** *8*(4), 253–262.

Kejnovsky, E.; Hawkins, J. S.; Feschotte, C. Plant Transposable Elements: Biology and Evolution. In *Molecular Biology and Evolution of the Plant Genome*; Wendel, J. F., Ed.; Springer Verlag: Vienna/New York, 2012; pp 17–34.

Keyte, A. L.; Percifield, R.; Liu, B.; Wendel, J. F. Intraspecific DNA Methylation Polymorphism in Cotton (Gossypium hirsutum L.). *J. Hered.* **2006,** *97*, 444–450.

Kronholm, I.; Collins, S. Epigenetic Mutations Can Both Help and Hinder Adaptive Evolution. *Mol. Ecol.* **2016,** *25*, 1856–1868.

Kucharski, R.; Maleszka, J.; Foret, S.; Maleszka, R. Nutritional Control of Reproductive Status in Honeybees via DNA Methylation. *Science* **2008,** *319*(5871), 1827–1830.

Lachmann, M.; Jabloka, E. The Inheritance of Phenotypes: An Adaptation to Fluctuating Environments. *J. Theor. Biol.* **1996,** *181*, 1–9.

Li, Q.; Eichten, S. R.; Hermanson, P. J.; Springer, N. M. Inheritance Patterns and Stability of DNA Methylation Variation in Maize Near-Isogenic Lines. *Genetics* **2014,** *196*(3), 667–676.

Martin, L. B.; Liebl, A. L.; Trotter, J. H.; Richards, C. L.; McCoy, K.; McCoy, M. W. Integrators: Physiological Determinants of Phenotypic Plasticity. *Integr. Comp. Biol.* **2011,** 51, 514–527.

Matzke, M. A.; Mosher, R. A. RNA-Directed DNA Methylation: An Epigenetic Pathway of Increasing Complexity. *Nat. Rev. Genet.* **2014,** *15*, 394–408.

Medrano, M.; Herrera, C. M.; Bazaga, P. Epigenetic Variation Predicts Regional and Local Intraspecific Functional Diversity in a Perennial Herb. *Mol. Ecol.* **2014,** *23*, 4926–4938.

Miklos, G. L.; Maleszka, R. Epigenomic Communication Systems in Humans and Honey Bees: from Molecules to Behavior. *Horm. Behav.* **2011,** *59*(3), 399–406.

Ossowski, S.; Schneeberger, K.; Lucas-Lledo, J. I.; Warthmann, N.; Clark, R. M.; Shaw, R. G.; et al. The Rate and Molecular Spectrum of Spontaneous Mutations in Arabidopsis thaliana. *Science* **2010,** *327*, 92–94.

Preite, V.; Verhoeven, K. J. F. Epigenetic Variation in Asexually Reproducing Organisms. Evolution **2014,** *68*, 644–655.

Rapp, R. A.; Wendel, J. F. Epigenetics and Plant Evolution. *New Phytol.* **2005,** *168*, 81–91.

Reinders, J.; Wulff, B. B.; Mirouze, M.; Mar_ı-Ord_o~nez, A.; Dapp, M.; Rozhon, W.; et al. Compromised Stability of DNA Methylation and Transposon Immobilization in Mosaic Arabidopsis Epigenomes. *Genes Dev.* **2009,** *23*, 939–950.

Richards, C. L.; Bossdorf, O.; Pigliucci, M. What Role Does Heritable Epigenetic Variation Play in Phenotypic Evolution? *Bioscience* **2010b,** *60*, 232–237.

Richards, C. L.; Schrey, A.; Pigliucci, M. Invasion of Diverse Habitats by Few Japanese Knotweed Genotypes is Correlated with Epigenetic Differentiation. *Ecol. Lett.* **15,** 1016–1025.

Richards, C. L.; Verhoeven, K. J. F.; Bossdorf, O. Evolutionary Significance of Epigenetic Variation. In *Molecular Biology and Evolution of the Plant Genome*; Wendel, J. F., Ed.; Springer Verlag: Vienna/New York, 2012a; pp 257–274

Richards, C. L.; Walls, R.; Bailey, J. P.; Parameswaran, R.; George, T.; Pigliucci, M. Plasticity in Salt Tolerance Traits Allows for Invasion of Salt Marshes by Japanese Knotweed s.l. (Fallopia japonica and F. _ bohemica, Polygonaceae). *Am. J. Bot.* **2008,** *95*, 931–942

Richards, E. J. Inherited Epigenetic Variation – Revisiting Soft Inheritance. *Nat. Rev. Genet.* **2006,** *7*, 395–401.

Richards, E. J. Natural Epigenetic Variation in Plant Species: A View from the Field. *Curr. Opin. Plant Biol.* **2011,** *14*, 204–209.
Roberts, R. J.; Vincze, T.; Posfai. J.; Macelis Roberts, R. J.; Vincze, T.; Posfai, J.; Macelis, D. REBASE Enzymes and Genes for DNA Restriction and Modification. *Nucleic Acids Res.* **2007,** 35(Database Issue), D269–D270.
Robertson, M. H.; Richards, C. L. Non-Genetic Inheritance in Evolutionary Theory – the Importance of Plant Studies. *Non-Gen. Inherit.* **2015,** *2*, 3–11.
Salmon, A.; Ainouche, M. L.; Wendel, J. F. Genetic and Epigenetic Consequences of Recent Hybridization and Polyploidy in Spartina (Poaceae). *Mol. Ecol.* **2005,** *14*, 1163–1175.
Salmon, A.; Clotault, J.; Jenczewski, E.; Chable, V.; Manzanares-Dauleux, M. Brassica oleracea Displays a High Level of DNA Methylation Polymorphism. *Plant Sci.* **2008,** *174*(1), 61–70.
Saze, H.; Tsugane, K.; Kanno, T.; Nishimura, T. DNA Methylation in Plants: Relationship to Small RNAs and Histone Modifications, and Functions in Transposon Inactivation. *Plant Cell Physiol.* **2012,** *53*(5), 766–784.
Schmitz, R. J.; Schultz, M. D.; Urich, M. A.; Nery, J. R.; Pelizzola, M.; Libiger, O.; et al. Patterns of Population Epigenomic Diversity. *Nature* **2013,** *495*, 193–198.
Schrey, A. W.; Alvarez, M.; Foust, C.; Kilvitis, H. J.; Lee, J. D.; Liebl, A. L.; Martin, L. B.; Richards, C. L.; Robertson, M. Ecological Epigenetics: Beyond MS-AFLP. *Integr. Comp. Biol.* **2013,** *53*(2), 340–350.
Schrey, A. W.; Coon, C.; Grispo, M.; Awad, M.; McCoy, E.; Mushinsky, H.; Richards, C. L.; Martin, L. B. Epigenetic Variation May Compensate for Decreased Genetic Variation with Introductions: A Case Study Using House Sparrows (Passer domesticus) on two Continents, 2012.
Scoville, A. G.; Barnett, L. L.; Bodbyl-Roels, S.; Kelly, J. K.; Hileman, L. C. Differential Regulation of a MYB Transcription Factor is Correlated with Transgenerational Epigenetic Inheritance of Trichome Density in Mimulus guttatus. *New Phytol.* **2011,** *191*, 251–263.
Skinner, M. K.; Manikkam, M.; Guerrero-Bosagna, C. Epigenetic Transgenerational Actions of Environmental Factors in Disease Etiology. *Trends Endocrinol. Metab.* **2010,** *21*(4), 214–222.
Slotkin, R. K.; Nuthikattu, S.; Jiang, N. The Evolutionary Impact of Transposable Elements on Gene and Genome Regulation. In *Molecular Biology and Evolution of the Plant Genome*; Wendel, J. F., Ed.; Springer Verlag: Vienna/New York, 2012.
Takeda, S.; Paszkowski, J. DNA Methylation and Epigenetic Inheritance During Plant Gametogenesis. *Chromosoma* **2006,** 115, 27–35.
Taudt, A.; Colom_e-Tatch_e, M.; Johannes, F. Genetic Sources of Population Epigenomic Variation. *Nat. Rev. Genet.* **2016,** *17*(6), 319–332.
Teixeira, F. K.; Heredia, F.; Sarazin, A.; Roudier, F.; Boccara, M.; Ciaudo, C.; et al. A Role for RNAi in the Selective Correction of DNA Methylation Defects. *Science* **2009,** *323*(5921), 1600–1604.
Tost, J. DNA Methylation: An Introduction to the Biology and the Disease-Associated Changes of a Promising Biomarker. *Mol. Biotechnol.* **2010,** *44*(1), 71–81.
Ungerer, M. C.; Johnson, L. C.; Herman, M. A. Ecological Genomics: Understanding Gene and Genome Function in the Natural Environment. *Heredity* **2008,** *100*(2), 178–183.

van der Graaf, A.; Wardenaar, R.; Neumann, D. A.; Taudt, A.; Shaw, R. G.; Jansen, R. C.; et al. Rate, Spectrum and Evolutionary Dynamics of Spontaneous Epimutations. *Proc. Natl. Acad. Sci. USA* **2015,** *112*, 6676–6681.

Vaughn, M. W.; Tanurdzic, M.; Lippman, Z.; Jiang, H.; Carrasquillo, R.; Rabinowicz, P. D.; Dedhia, N.; McCombie, W. R.; Agier, N.; Bulski, A.; Colot, V.; Doerge, R. W.; Martienssen, R. A. Epigenetic Natural Variation in Arabidopsis thaliana. *PLoS Biol.* **2007,** *5*, 1617–1629.

Verhoeven, K. J. F.; Jansen, J. J.; van Dijk, P. J.; Biere, A. Stress-induced DNA Methylation Changes and their Heritability in Asexual Dandelions. *New Phytol.* **2010,** *185*(4), 1108–1118.

Weaver, I. C. G.; Cervoni, N.; Champagne, F. A.; D'Alessio, A. C.; Sharma, S.; Seckl, J. R.; Dymov, S.; Szyf, M.; Meaney, M. J. Epigenetic Programming by Maternal Behavior. *Nat Neurosci.* **2004,** *7*(8), 847–854.

West-Eberhard, M. J. *Developmental Plasticity and Evolution*; Oxford University Press: New York, 2003.

CHAPTER 6

QTL Analysis and Reverse Genetics in Molecular Systematics

SUMIAH WANI, SHAHJAHAN RASHID, FARHANA WANI, and AFLAQ HAMID*

Division of Plant Pathology, Sher-e-Kashmir University of Agricultural Science and Technology, Shalimar Srinagar, Kashmir, India

*Corresponding author. E-mail: falak19@gmail.com

ABSTRACT

Molecular systematics, also known as molecular biology, molecular evolution, and quantitative systematics, is a hybrid between conventional systematics and the rapidly growing field of bioinformatics. Actually, molecular systematics is a method that uses molecular composition evidence from living species, such as DNA, RNA, or both, to answer questions about their correct classification from a taxonomy perspective in evolutionary biology, rather than relying solely on morphology. Since DNA, RNA, and protein sequences are highly heritable units, they are more suitable for systematic studies than morphological characters, which are predisposed to many environmental factors. Furthermore, molecular characters are typically simpler to grasp than morphological characters, and the evolution of molecular features is more regular than that of morphological features, resulting in a simple and solid picture of relationships. Molecular characters are more numerous than morphological characters, and a large number of them can be made in a limited amount of time. Two important areas that

Plant Ecogenomics. Peerzada Arshid Shabir, PhD, Peerzada Yasir Yousuf, PhD & Khalid Rehman Hakeem, PhD (Eds.)

apply to molecular systematics are quantitative trait loci (QTL) research and reverse genetics. QTLs are specific DNA segments that control the expression of a quantitative phenotype. The primary aim of QTL research is to genotype a large number of individuals with significant variation in the trait(s) of interest across multiple loci using a set of markers that spans the majority of the genome. Furthermore, reverse genetics, a new technique that is a direct result of recombinant DNA science, operates in the opposite way, beginning with DNA or a protein about which no prior genetic knowledge existed and acting backward to the gene before finally arriving at the phenotype.

6.1 SYSTEMATICS

Systematics is a branch of science that studies biological diversity and organizes it into a classification system. The first classification was discovered in Ancient Greece, particularly with Aristotle, and Linnaeus shaped the hierarchical classification scheme that is still used in taxonomy in 1758. Biological variation began to be accounted as the result of species diverging from a single ancestor with the advent of Darwin's theory of evolution in the 19th century. Molecular systematics (biochemical taxonomy) is a method of estimating the number of variations between organisms by comparing functionally homologous molecular sequences from each organism being examined. The species with the most variations are more likely to be distantly related. Furthermore, since the number of equivalent amino acids substituted by nucleotide substitutions is usually proportional to time, there is evidence of the time scale involved. This knowledge has mostly been useful in areas where the fossil record is lacking, and it can be paired with other anatomical, physiological, and embryological data to produce more detailed phylogenetic trees. Charls G. Sibley applied molecular systematics to birds, Morris Goodman to primates, and Herbest C. Dessauer to reptiles, followed by Allan C. Wilson, Robert K. Sulender, and John, who used their work with protein electrophoresis, which began around 1956. Botanists have used morphology for plant recognition and classification for hundreds of years. Molecular systematics has also played an important part in understanding the fundamentals of plant relationships and evolution. It does, however, have drawbacks. Plant homology, which means that two identical constructs share the same evolutionary origin, is considered one of the most important restrictions on morphology. In other words, it suggests that a trait was inherited by an ancestor and passed on over the generations. However, homology in plant morphology continues to be difficult to overcome since plant architectures can alter and take on new shapes (e.g., leaves are modified to spines in cacti). Botanists compare the shape of leaves to distinguish between two plants, and molecular systematists do the same, but in two cases, molecules have an edge over morphology. Molecular homology is usually much easier to discover for beginners than morphology. Second, molecules provide an abundance of information that morphology alone cannot provide (Nair, 2019). A scientist studying morphology compares 100 traits, while a scientist studying molecules compares hundreds of thousands of traits, based on the methods used. Molecular systematics began with micro molecules in earlier days, and the first of these experiments can be traced back to the 1880s, although there was a lot of work done between the 1950s and 1970s. Plants' colors, scents, and chemical defences are all regulated by micro molecules, which are small molecules. The chemicals found in different plants are first identified and compared to see whether there are any correlations between species. Similar compound-sharing species are considered to be more closely related. Botanists later pioneered the use of macromolecules, such as proteins and nucleic acids. The majority of research on proteins took place in the 1970s, with the aim of determining if different organisms or groups of plants had different types of specific enzymes (protein sequencing) or determining whether different organisms or groups of plants had different types of specific enzymes (isozyme variability). Other protein-related experiments used serological concepts to create antibodies against protein samples, which were then compared to extracts from various organisms. The degree to which one plant's antibodies complement the proteins of another plant can be used to determine their relationship (Nair, 2019). The true strength of molecular systematics lies in analyzing variations in species over evolutionary time, as well as interactions between species with no physical characteristics in general. The rate of transition in DNA over time serves as a molecular clock, showing how much evolutionary time has elapsed. In the late 1960s and 1970s, DNA–DNA hybridization was used to study the application of deoxyribonucleic acid (DNA). The theory of "Double-stranded structure of DNA molecule" is used in this system, which states that high temperatures (greater than 80°C) induce denaturation of all DNA, making it single-stranded. As DNA is cooled, it reverts to a double-stranded structure (re-annealing) and the temperature at which the DNA strands become fully double-stranded shows how close the strands are (Nair, 2019). The DNA of two different plants is mixed and heated in this process. If all of their DNA comes from closely related plants, the re-annealing temperature is high, but it is lower if it comes from two distantly related plants. As a consequence, the re-annealing temperature is an estimate of plant similarity. In the 1980s, botanists used restriction site analysis to compare DNA between plants. Restriction enzymes were used to cut DNA into different length fragments. When a particular sequence of nucleotides is encountered, these enzymes are able to split the DNA at the specific combination of nucleotides any time. Size on a gel electrophoresis isolates the fragments, which are then visualized with a probe that matches specific DNA regions. By studying these fragments, it is possible to establish whether a specific restriction site is present or absent in any given species. When two or more plants share a restriction site, it means they share a more recent shared ancestor. Since the chloroplast DNA is small in contrast to other sections of the genome and a variety of probes are available, botanists often use it (Nair, 2019).

6.2 MOLECULAR METHODS IN SYSTEMATICS

The polymerase chain reaction (PCR), developed by Mullis and colleagues, revolutionized the molecular biology methodological arsenal (Saiki et al., 1988; Mullis et al., 1994). This approach outperforms molecular cloning because it amplifies the DNA sequence of interest to large copy numbers in vitro. Two single-stranded oligonucleotide primers are needed for amplification of a particular DNA sequence that is complementary to motifs on the template DNA. The primer sequences have been chosen to allow base-specific binding to the two reverse-oriented template stands. With the development of emerging technology, a wide range of molecular markers is now available (Neir, 2019). The following are brief overviews of several important and appropriate molecular markers that are commonly used in systematics, genetic variation, and phylogenetic studies.

6.2.1 PROTEIN MARKERS

Protein markers are the earliest recognized molecular markers focused on protein polymorphisms, and the electrophoretic isolation of proteins, accompanied by precise staining of distinct protein subgroups, is a commonly used procedure. Several experiments have focused on enzyme effectiveness using enzymatic and non-enzymatic proteins, as well as genetic diversity in natural populations, gene flow, species identification, hybridization, and phylogenetic relationships (Murphy et al., 1996). After electrophoretic separation, proteins are visualized on polyacrylamide gels with a universal protein stain (Coomassie Brilliant Blue), and allelic differences are observed as variations in polypeptide size (Burow and Blake 1998). Using electrophoretic techniques, two types of protein data may be received at the same time. One that is derived from isozymes, which are all chemically related enzyme types (Markert and Moller 1959). The other collection of data includes allozymes, which are a subclass of isozymes that are polypeptide variants that display various alternate alleles of the same gene locus. In molecular systematics, both types of data are important because they include proteins that can be differentiated based on their net charge and size (Murphy et al., 1996). Codominant succession, cost effectiveness, and technological ease are the advantages of protein markers. However, there are a host of drawbacks. The main drawback of this approach is that it does not yield high levels of polymorphisms, which are needed to classify a large number of genotypes in a population (Cruzan, 1998), since enzyme activity is necessary to research a small number of loci (Beebee and Rowe, 2004). As a result, DNA-based approaches often replace or act in combination with allozymes in order to generate more polymorphisms.

6.2.2 RESTRICTION FRAGMENT LENGTH POLYMORPHISM (RFLP)

The move from protein markers to DNA markers in genetic marker research occurred in the early 1980s, owing to the increased number of possible markers available. The restriction fragment length polymorphism (RFLP) method was first used to build a genetic map of humans (Botstein et al., 1980). Using restriction endonucleases that recognize unique four to six base-pair palindromic sequences, genomic DNA is first digested in this process. The fragments are then electrophoretically isolated on agarose gels by bulk, immobilized on nylon or nitrocellulose membranes after being denatured. The use of radiolabeled probes allows for the visualization of specific DNA fragments. The blots are then seen by X-ray film to reveal markers (Isaac, 1994). A difference in restriction enzyme sites is the cause of RFLP-detected polymorphism. RFLP markers are typically locus specific and codominant. While RFLP analysis is useful for important experiments, it is a labor-intensive and time-consuming method. Other disadvantages include the need for high-quality, large-amount DNA and DNA probes for hybridization. Cleaved amplified polymorphic sequence (CAPS), also called as PCR-RFLP, is a type of RFLP marker that has been modified (Williams et al., 1991; Konieczny and Ausubel, 1993). Unlike standard RFLP analysis, the CAPS procedure can be done with a limited amount of DNA template and the use of dangerous radioactive probes or blotting steps is not required. However, to assess suitable combinations of amplicons and restriction enzymes, a wide range of PCR primer pairs and restriction enzymes must be checked during the initial step of a CAPS analysis (Nair, 2019).

6.2.3 RANDOM AMPLIFICATION OF POLYMORPHIC DNA (RAPD)

Random amplification of polymorphic DNA (RAPD) (Williams et al., 1990) is the most popular DNA fingerprinting method, in which short about 10 base primers of any sequence are annealed for direct amplification of items of unknown sequence. RAPD uses just one primer, unlike other PCR-based systems. Since the primers have a short chain, there are more possible primer-binding sites in the genome, and where two primer-binding sites are near together, the likelihood of successful replication of DNA fragments improves. Separation of amplified DNA on agarose gels is performed electrophoretically, after that stained with ethidium bromide and examined under ultraviolet light. The RAPD technique is technically effective, simple, and allows for high throughput (Deragon and Landry, 1992), resulting in widespread acceptance among scientists. The use of short oligonucleotide primers of any sequence for PCR amplification of genomic regions offers an almost never-ending source of insightful genetic polymorphisms (Williams et al., 1990; Welsh and McClelland, 1990). The PCR's sensitivity in DNA amplification, therefore, decreases the need for significant volumes of target genomic DNA (Saiki et al., 1988). Environmental effects have little impact on specific molecular genetic profiles, which gives RAPD an edge over morphological measurements and isozyme study. RAPDs are dominant markers, which has a drawback in genetic studies because they cannot diagnose heterozygosity as well as codominant marker systems do. These RAPD bands are isolated on a gel according to their molecular weight (fragment size) and electrical charge.

6.2.4 MICROSATELLITE DNA MARKERS SIMPLE SEQUENCE REPEATS

Microsatellite DNA are DNA repeats with tandem repeats of a central sequence of two to five base pairs, also known as simple sequence repeats (SSRs). The number of tandem repeats in the core sequence varies between alleles. These sequences are present in the eukaryotic genome, and thousands of possible SSR markers can be designed for most animals. In addition, SSR markers have a lot of influence when it comes to determining diversity. Since these genes are subject to high mutation rates and their expression is codominant, it is possible to distinguish homozygotes from homozygotes with greater accuracy (Leberg, 1992). However, since the production of microsatellite markers for an increasing variety of species is accelerating, it is likely that microsatellite development will become less expensive in the immediate future (Nair, 2019).

6.2.5 INTER-SIMPLE SEQUENCE REPEATS

The inter-simple sequence repeats (ISSR) marker, like RAPD, does not need any previous knowledge of the genome sequence (Godwin et al., 1997). ISSR uses primers attached at the 5′ or 3′ end of a repeat region and expanded into the flanking region to track variability in different microsatellite regions. The genomic fragments can then be amplified between inversely directed repeats using this technique (ISSRs). The study of the ISSR marker entails three steps: (1) design of oligonucleotide primers, (2) PCR amplification of ISSR fragments, and (3) PCR-product separation ISSR is a multi-locus marker that can be used for fingerprinting, genetic diversity research, and genetic imaging. This technique is fast and can differentiate between people who are closely related. ISSR markers provide a variety of advantages, including polymorphic loci, high throughput, and low cost. ISSRs, including RAPDs, look for variation in the many microsatellite regions found in the genome. ISSR markers were first used in experiments on cultivated species (Zietkiewicz et al., 1994; Gupta et al., 1994), showing the hypervariable existence of ISSR makers (Wolfe and Liston, 1998; Wolfe et al., 1998a, 1998b). ISSR markers have a high variability and "mapping density" as compared to RFLP and RAPD data, making them perfect for generating genetic maps of individual organisms (Nagaoka and Ogihara, 1997). They are superior to other marker systems available for investigating genetic variation among closely related individuals and in crop cultivar classification because they are more stable in repeatability experiments and less resistant to modifying band patterns with changes in template DNA concentrations (Fang and Roose, 1997; Nagaoka and Ogihara, 1997). ISSR markers have the downside of being dominant and less repeatable.

6.2.6 AMPLIFIED FRAGMENT LENGTH POLYMORPHISM (AFLP)

Vos et al. were the first to identify amplified fragment length polymorphism (AFLP) markers in 1995. Absolute digestion of genomic DNA is accompanied by partial PCR amplification of restriction fragments in this technique. Three steps are included in this technique: DNA restriction and oligonucleotide adapter ligation, selective amplification of sets of restriction fragments, and agarose gel analysis of the amplified fragments. Adapters and the restriction site series are used as reference sites for primer annealing in PCR amplification of restriction fragments. Primers that fit the nucleotides flanking the restriction sites are used to achieve selective amplification of restricted fragments. PCR may be used to visualize sets of restriction fragments using this approach without understanding the nucleotide sequence (Fig. 6.2). AFLP may be used to investigate varietal recognition, biological variation, and genetic mapping (Zhu et al., 1998).

6.2.7 INTERNAL TRANSCRIBED SPACER (ITS)

The ribosomal DNA cistron of nuclear ribosomes is one of the regions that has been widely used in phylogenetic studies of various species (Hillis and

Dixon, 1991). Nuclear genes that are grouped into transcriptional rDNA units with a small 18S/SSU (smaller sub unit), a 5.8S, and a wide and ITS2 encode cytoplasmic ribosomal RNAs (rRNA) in most eukaryotes (Fig. 6.1). Since the DNA sequences for larger sub unit (LSU) and smaller sub unit (SSU) rDNA play such an important role in ribosome synthesis, they are subject to good stabilizing selection (Nair, 2019). The non-coding ITS rDNA regions are not subject to the same technical limitations as the coding regions. Point mutations account for the majority of the difference in ITS sequences. As a result of the rapid aggregation of mutations, these regions are extremely variable. Phylogenetically analyzing the composite data sets from both the ITS1 and ITS2 spacers yields trees with better resolution and internal protection than any spacer alone. The beneficial effect of simultaneous analysis is not unexpected given the low number of usable characters in each spacer. As a result, ITS regions have been extensively used to investigate the spectrum of interfamily to intraspecies genetic differences in a variety of plant taxa (Bakker et al., 1992; Connell, 2000; Lundholm et al., 2006).

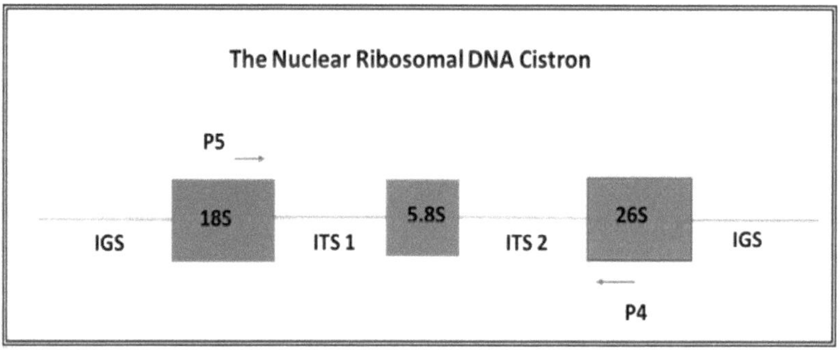

FIGURE 6.1 The nuclear ribosomal DNA cistron.

6.2.8 SINGLE NUCLEOTIDE POLYMORPHISMS (SNPS)

Single nucleotide polymorphisms (SNPs) are variations of DNA sequence caused by the modification of a single nucleotide in the genetic code leading to a gene, a part of a gene, or even a strip of DNA being modified. SNPs must exist in at least 1% of the overall population to be considered for these variants. SNPs are the most frequent form of DNA sequence variation, and they've been used in animal and human genetics research, including the mapping of the human genome (and so on.) (Wang et al., 1998). In contrast to other genetic markers, SNPs are more widespread and conserved throughout the genome (Osman et al., 2003). SNPs have been used widely in medical science during the past decade (Chen and Sullivan, 2003). SNP determination can influence the accuracy of predicting genetic occurrences within populations, according to biologists who tested the capacity of SNPs to estimate population parameters (Kuhner et al., 2000). SNPs have a wide variety of uses, from researching human and cattle species to bacteria and plants. Since SNPs are yet to be used to infer phylogenetic relationships in plants, the discovery of SNP-based markers within a taxon would be useful in phylogenetic analysis.

6.3 QTL ANALYSIS

We have discussed about methods that start with variation in DNA expression (molecular variation) up to this stage. We will explore how phenotypic variance can be used as a starting point for studying the molecular ecology of adaptive traits in this section. Many agriculturally significant traits, such as yield, efficiency, and certain types of disease tolerance, are now understood to be quantitative traits/polygenic/multifactorial or complex traits, as they are influenced by a large number of genes. Quantitative trait loci (QTL) are genes within genomes that are correlated with a specific QTL. In particular, microarray studies show that the expression of several phenotypes is mediated by networks of hundreds of genes (Freeland et al., 2011). Even, it's useful to know how many genetic loci affect the expression of a single trait, where these loci are in the genome in relation to other loci, and which of them accounts for the majority of phenotypic variation seen in nature (Collard et al., 2005). It is not possible to identify QTLs solely by traditional phenotypic assessment. In the 1980s, the identification of molecular markers ushered in a major revolution in the classification of quantitative traits, allowing for the compilation of QTLs. QTLs are DNA regions that control how a quantitative phenotype is expressed. And the aim of QTL research is to genotype a large population that is substantially variable in the trait(s) of interest using a series of markers spread over multiple loci that cover a large portion of the genome. This research can be conducted on wild species or on crosses that have a proven pedigree and are produced under stable conditions (Joanna et al., 2011). Examination of variance (ANOVA) is the most fundamental approach for identifying markers that describe a large amount of phenotypic heterogeneity. Those markers can be compared to the gene that regulates the quantitative trait's expression. As previously reported, QTL research involves the generation of a large number of markers in order to provide extensive coverage of the entire genome. Because of their random distribution across the genome, RFLPs and AFLPs were once thought to be effective markers. RFLPs, in contrast, are not commonly used, and one of the main drawbacks of AFLPs is that they are dominant, unable to differentiate homozygotes from heterozygotes. SSRs are often used for QTL analysis, but generating a large enough number of SSRs to cover the whole genome is difficult. SNPs have become the chosen marker class for QTL studies as new strategies have emerged, as they are co-dominant markers with broad coverage of the genome. In certain instances, a mixture of different markers is recommended for QTL study.

6.4 REVERSE GENETICS

Gregor Mendel, an Augustinian monk, pioneered the study of "Genetics" in the 1860s, when he conducted studies that indicated the presence of chromosomes (Griffiths et al., 2000). Later, in the 1950s, James Watson and Francis Crick uncovered the DNA structure, and in the 1980s, Kary Mullis' polymerase chain reaction (PCR) technique (Saiki et al., 1985; Mullis and Faloona, 1987) sparked a genetic revolution. Technological advances in sequencing have increased the accumulation of genetic sequence data to the point that entire genome genomes for numerous species, including pathogens, are now freely accessible (Tierney and Lamour, 2005). The area of genetic science is increasingly expanding, with reverse genetics being one of the genetic research methods that can be used to analyze gene function. The traditional genetics approach starts with the mutant allele, then uses progeny ratio analysis to find the identity of the associated gene, and eventually sequences the gene to determine its DNA and protein sequence. Instead, reverse genetics, a modern approach to recombinant DNA science, operates in the opposite way (Fig. 6.2), starting with a protein or DNA about which no genetic information is available and working backward to create a mutated gene and eventually a mutant phenotype associated with it.

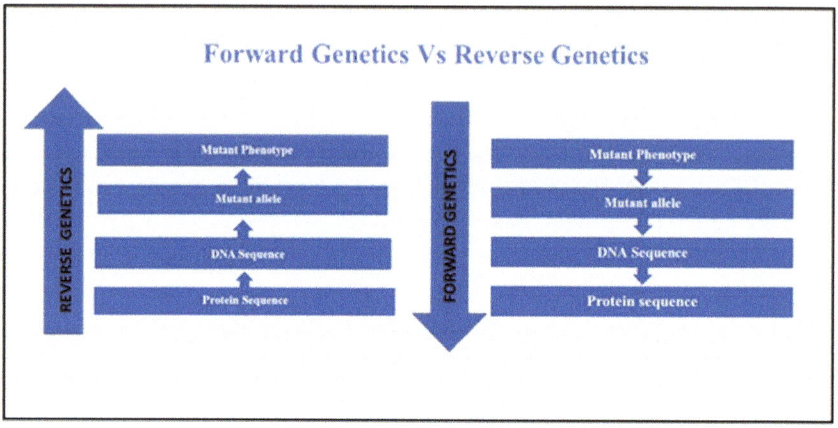

FIGURE 6.2 Forward genetics vs. reverse genetics.

To validate the action of individual genes as well as their properties, physiologists and molecular biologists have been inactivating particular genes. The "knockout" mouse model, in which genes are inactivated primarily for human medical studies, is the most famous example of this (Joanna et al., 2011). The primary goal of inactivating a gene is to identify phenotypic variations between species that have both active and inactive copies of the target gene. This enables the identification of the particular protein encoded by the target gene at the molecular level, as well as the corresponding phenotypic and/or fitness significance. Gene inactivation can be accomplished in a variety of ways. It can either be knocked out using genetic modification techniques by injecting a foreign DNA sequence into the target gene, making it non-functional, or certain groups of RNAs can help control gene transcription by a mechanism known as RNA interference (RNAi). Small interfering RNAs (siRNAs) are 20–25 base pair double-stranded RNA molecules that inhibit the expression of a particular gene. MicroRNAs (miRNAs) are single-stranded sequences of around 21–23 base pairs in length that bind to complementary miRNAs to prevent particular proteins from being translated. These miRNAs, however, unlike siRNAs, are not very precise and can target hundreds of different mRNAs. With the aid of a vector, such as a virus, these small RNAs may be inserted into a target organism. In studies of organismal ecology, evolution, and phylogenetic studies, molecular ecologists have recently begun to use gene inactivation approaches (Joanna et al., 2011). Zavala et al. silenced a significant anti-herbivore defence gene in the tobacco plant *Nicotiana attenuata* in 2004 and demonstrated that inactivating the gene enabled the plant to develop far more seed capsules than the wild form. In this plant genus, this experiment revealed a direct trade-off between reproductive fitness and anti-herbivore defence. To use gene silencing in molecular systematics in conjunction with other genomic techniques, many other model systems are currently being developed (Freeland et al., 2011).

KEYWORDS

- **molecular systematics**
- **QTL**
- **reverse genetics**

REFERENCES

Bakker, F. T.; Olsen, J. L.; Stam, W. T.; van den Hoek, C. Nuclear Ribosomal DNA Internal Transcribed Spacer Regions (ITS1 and ITS2) Define Discrete Biogeographic Groups in Cladophora albida (Chlorophyta). *J. Phycol.* **1992,** *28*, 839–845.

Beebee, T.; Rowe, G. *An Introduction to Molecular Ecology*; Oxford University Press: Oxford, 2004.

Blazej, R. G.; Paegel, B. M.; Mathies, R. A. Polymorphism Ratio Sequencing: A New Approach for Single Nucleotide Polymorphism Discovery and Genotyping. *Gen. Res.* **2003,** *13*, 287–293.

Botstein, D.; White, R. L.; Skolnick, M.; Davis, R. W. Construction of a Genetic Linkage Map in Man Using Restriction Fragment Length Polymorphism. *Am. J. Hum. Gen.* **1980,** *32*, 314–331.

Brookes, A. J. The Essence of SNPs. *Gene* **1999,** *234*, 177–186.

Burow, M. D.; Blake, T. K. Molecular Tools for the Study of Complex Traits. In: *Molecular Dissection of Complex Traits*; Paterson, A. H., Ed.; CRC Press: Washington, DC, 1998; pp. 13–29.

Chen, X.; Sullivan, P. F. Single Nucleotide Polymorphism Genotyping: Biochemistry, Protocol, Cost and Throughput. *Pharmacogenomics J* **2003,** *3*, 77–96.

Collard, B.C.; Jahufer, M.Z.Z.; Brouwer, J.B.; Pang, E.C.K. An Introduction to Markers, Quantitative Trait Loci (QTL) Mapping and Marker-Assisted Selection for Crop Improvement: The Basic Concepts. *Euphytica* **2005,** *142* (1), 169–196.

Connell, L. B. Nuclear ITS Region of the Alga Heterosigma akashiwo (Chromophyta, Raphidophyceae) Is Identical in Isolates from Atlantic and Pacific Basins. *Marine Biol.* **2000,** *136* (6), 953–960.

Cruzan, M. Genetic Markers in Plant Evolutionary Ecology. *Ecology* **1998,** *79*, 400–412.
Deragon, J. M.; Landry, B. S. RAPD and Other PCR-Based Analysis of Plant Genomes Using DNA Extracted from Small Leaf Discs. *PCR Methods Appl.* **1992,** *1*, 175–180.
Fang, D. Q.; Roose, M. L. Identification of Closely Related Citrus Cultivars with Inter-Simple Sequence Repeat Markers. *Theor. App. Gen.* **1997,** *95*, 408–417.
Freeland, J. R.; Kirk, H.; Petersen, S. Studying Ecologically Important Traits: Ecogenomics, QTL Analysis, and Reverse Genetics. *Mol. Ecol.* **2011**. doi:10.1002/9780470979365.ch5
Godwin, I. D.; Aitken, E. A. B.; Smith, L. W. Application of Inter Simple Sequence Repeat (ISSR) Markers to Plant Genetics. *Electrophoresis* **1997,** *18*, 1524–1528.
Gupta, M.; Chyi, Y. S.; Romero-Severson, J.; Owen, J. L. Amplification of DNA Markers from Evolutionary Diverse Genomes Using Single Primers of Simple-Sequence Repeats. *Theor. App. Gen.* **1994,** *89*, 998–1006.
Isaac, P. G. *Methods in Molecular Biology: Protocols for Nucleus Acid Analysis by Nonradioactive Probes*; Humana Press Inc.: Totowa, NJ, 1994.
Jaccoud, D.; Peng, K.; Feinstein, D.; Kilian, A. Diversity Arrays: A Solid State Technology for Sequence Information Independent Genotyping. *Nucl. Acids Res.* **2001,** *29*, 25.
Joanna, R.; Heather, F.; Kirk, R. H.; Petersen, S. Studying Ecologically Important Traits: Ecogenomics, QTL Analysis, and Reverse Genetics, **2011**. https://doi.org/10.1002/9780470979365.ch5
Konieczny, A.; Ausubel, F. M. A Procedure for Mapping Arabidopsis Mutations Using Co-Dominant Ecotype-Specific PCR-Based Markers. *Plant J.* **1993,** *4*, 403–410.
Kuhner, M. K.; Beerli, P.; Yamato, J.; Felsenstein, J. Usefulness of Single Nucleotide Polymorphism Data for Estimating Population Parameters. *Genetics* **2000,** *156*, 439–447.
Leberg, P. L. Effects of Population Bottlenecks on Genetic Diversity as Measured by Allozyme Electrophoresis. *Evolution* **1992,** *46*, 477–494.
Lundholm, N.; Moestrup, O.; Kotaki, Y.; Hoef Emden, K.; Scholin, C.; Miller, P. Inter- and Intraspecific Variation of the Pseudo nitzschia delicatissima Complex (Bacillariophyceae) Illustrated by rRNA Probes, Morphological Data and Phylogenetic Analyses. *J Phycol.* **2006,** *42*, 464–481.
Markert, C. L.; Moller, F. Multiple Forms of Enzymes, Tissue, Ontogenetic and Species Specific Pattern. *Proc. Natl. Acad. Sci. USA* **1959,** *45*, 753–763.
Murphy, R. W.; Sites, J. W. Jr.; Buth, D. G.; Haufler, C. H. Proteins: Isozyme Electrophoresis. In *Molecular Systematics*; Hillis, D. M., Moritz, C., Mable, B. K., Eds., 2nd ed.; Sinauer Associates: Sunderland, MA, 1996; pp 51–120.
Nair, K. N. Methods and Approaches in Plant Molecular Systematics, 2019.
Osman, A.; Jordan, B.; Lessard, P. A.; Muhammad, N.; Haron, M. R.; Riffin, N. M.; Sinskey, A. J.; Rha, C.; Housman, D. E. Genetic Diversity of Eurycoma longifolia Inferred From Single Nucleotide Polymorphisms. *Plant Physiol.* **2003,** *131*, 1294–1301.
Saiki, R. K.; Gelfand, D. H.; Stoffel, S.; Scharf, S. J.; Higuchi, R.; Horn, G. T.; Mullis, K. B.; Erlich, H. A. Primer Directed Enzymatic Amplification of DNA with a Thermostable DNA Polymerase. *Science* **1988,** *239*, 487–491.
Tierney, M. B.; Lamour, K. H. An Introduction to Reverse Genetic Tools for Investigating Gene Function. *Plant Health Instruct.* **2005,** *10*.
Wang, D. G.; Fan, J.-B.; Siao, C.-J.; Berno, A.; Young, P.; Sapolsky, R.; Ghandaur, G.; Perkins, N.; Winchester, E.; Spencer, J.; Kruglyak, L.; Stein, L.; Hsie, L.; Topaloglou, T.; Hubbell, E.; Robinson, E.; Mittmann, M.; Morris, M. S.; Shen, N.; Kilburn, D.; Rioux,

J.; Nusbaum, C.; Rozen, S.; Hudson, T. J.; Lipshutz, R.; Chee, M.; Lander, E. S. Large Scale Identification, Mapping and Genotyping of Single Nucleotide Polymorphisms in the Human Genome. *Science* **1998,** *280*, 1077–1082.

Welsh, J.; McClelland, M. Fingerprinting Genomes Using PCR with Arbitrary Primers. *Nucl. Acids Res.* **1990,** *18*, 7213–7218.

Williams, J. G. K.; Kubelik, A. R.; Livak, K. J.; Rafalski, J. A.; Tingey, S. V. DNA Polymorphisms Amplified Byarbitrary Primers Are Useful as Genetic Markers. *Nucleic Acids Res* **1990,** *18*, 6231–6235.

Williams, N. M. V.; Pande, N.; Nair, S.; Mohan, M.; Bennet, J. Restriction Fragment Length Polymorphism Analysis of Polymerase Chain Reaction Products Amplified from Mapped Loci of Rice (*Oryza sativa L.*) Genomic DNA. *Theor. Appl. Gen.* **1991,** *82*, 489–498.

Wolfe, A. D.; Liston, A. Contributions of PCR-Based Methods to Plant Systematics and Evolutionary Biology. In *Molecular Systematics of Plants II. DNA Sequencing*; Soltis, D. E., Soltis, P. S., Doyle, J. J., Eds.; Kluwer: Dordrecht, 1998; pp 43–86.

Wolfe, A. D.; Xiang, Q-Y.; Kephart, S. R. Assessing Hybridization in Natural Populations of Penstemon (Scrophulariaceae) Using Hypervariable Inter Simple Sequence Repeat (ISSR) Bands. *Mol. Ecol.* **1998a,** *7*, 1107–1125.

Wolfe, A. D.; Xiang, Q-Y.; Kephart, S. R. Diploid Hybrid Speciation in Penstemon (Scophulariaceae). *Proc. Natl. Acad. Sci. USA* **1998b,** *95*, 5112–5115.

Zhu, J.; Gale, M. D.; Quarrie, S.; Jackson, M. T.; Bryan, G. J. AFLP Markers for the Study of Rice Biodiversity. *Theor. Appl. Gen.* **1998,** *96*, 602–611.

Zhu, Y. L.; Song, Q. J.; Hyten, D. L.; Van Tassell, C. P.; Matukumalli, L. K.; Grimm, D. R.; Hyatt, S. M.; Fickus, E. W.; Young, N. D.; Cregan, P. B. Single Nucleotide Polymorphisms (SNPs) in Soybean. *Genetics* **2003,** *163*, 1123–1134

Zietkiewicz, E.; Rafalski, A.; Labuda, D. Genome Fingerprinting by Simple Sequence Repeat (SSR)- Anchored Polymerase Chain Reaction Amplification. *Genomics* **1994,** *20*, 176–183.

CHAPTER 7

Dispersal, Gene Flow, and Population Structure

NAVEED UL MUSHTAQ, SEERAT SALEEM, AADIL RASOOL, WASIFA HAFIZ SHAH, and REIAZ UL REHMAN*

Department of Bioresources, School of Biological Sciences, University of Kashmir, Srinagar 190006, India

**Corresponding authors. E-mail: rreiazbiores@gmail.com*

ABSTRACT

Genetic diversity among the populations is dependent on gene flow. The allele frequencies get modified leading to changes in the distribution of genetic diversity. Many factors affect the gene flow between populations. The species with lower dispersal, fragmented habitat, longer geographic distances, and smaller size tend to have low gene flow. Mobility plays an important role in population size as the pollen and seeds are dispersed long distances by different means/mediums. However, many isolated places have a lower rate of gene flow due to geographical isolation and minimal population size because it does not allow the exchange of genetic material. This leads to the development of genetic differences between the same species as well as a decreased population, which results in an increased divergence that is due to genetic drift. Similarly, new species arise from the same ancestral species due to reproductive barriers, which could be due to phenotypic differences, environmental conditions, fragmentation, and availability

Plant Ecogenomics. Peerzada Arshid Shabir, PhD, Peerzada Yasir Yousuf, PhD & Khalid Rehman Hakeem, PhD (Eds.)

of specialist pollinators. In this chapter, we shall discuss the dispersal modes, quantification of dispersal, bottleneck effect on gene flow, measurements of gene flow, estimation of gene flow, and the population structure. The currently available methods and models to study the population structure will be thoroughly examined.

7.1 INTRODUCTION

Gene flow or gene migration means the flow of genes from one population to the other. It can be also defined as the transmission of genetic material from one population to the other (Templeton, 2019). Movement of genes can take place within the same kind in a species by vertical gene transfer like from parents to offspring's or it can take place between different species by horizontal gene transfer like in prokaryotes (Wang et al., 2021). Flow-within-population will likely increase genetic disparity whereas in the case of distant populations it will decrease the genetic difference between the populations. This movement of genes is the foundation for evolution. It includes active or passive methods of movement like that of plants, animals, gametes, or seeds, etc. Gene flow is influenced by gender and limited to some stages of the life cycle (Bolstad et al., 2017). It can be fast under some climatic settings and reduced under others. It can also be affected by environmental conditions and different types of stresses, for example, pollen and grain development are receptive to temperature stress (Singh et al., 2012). At suitable intensities, gene flow is likely to homogenize populations and opposes drift.

Plants are habitually static organisms. Pollen, vegetative propagation, and dispersion of seeds are the main ways of gene flow (Garcia et al., 2007). Gene movement by pollen grains is cardinal for genetic variability in and between plant populations (Imbert and Lefèvre, 2003). Seed dispersal is also a key factor for the establishment of new sites and movement between neighboring subpopulations (Cain et al., 2000). A controlled flow of genes results in the spatial genomic structure of natural plant populations (Turner et al., 1982). The information about genetic structure within populations allows us to achieve insights associated with the dispersal behaviors of genetic materials and individuals (Rousset, 2000). In the case of anemophily, extensive array and undirected pollen dispersal are seen, but in animals, pollinated flora gene flow is controlled and depends on activities of the animal involved (García et al., 2007).

7.2 BARRIERS TO GENE EXCHANGE

Gene flow is barred before and after fertilization, that is why it is called prezygotic and postzygotic reproductive isolation, respectively (Hollinger and Hermisson, 2017). Reproductive isolation prevents gene flow and maintains separate species. Thus, these obstacles can form the structure of the population (Donnelly and Townson, 2000). The prezygotic barrier to gene flow takes place before the zygote is formed (species having a sexual mode of reproduction) (Nosil, 2012), for example, the difference in the signals given by different species and different courtship displays preferred by different species constitute premating prezygotic isolation. The difference in place, way, and time of mating in individuals from different species also comes under the purview of prezygotic isolation. Different preferences in the mode of pollination, for example, ornithophily and entomophily can also lead to premating prezygotic isolation (Ramsey et al., 2003). Postzygotic obstructions operate after hybrid zygotes are formed; these barriers for taking place when progeny from hybrid mattings are infertile. Genomic incompatibilities, which are not dependent on the environment, also contribute to postzygotic isolation (April et al., 2011). Prezygotic barriers include temporal, behavioral, mechanical, gametic, and habitat isolations. Behavioral or ethological isolation occurs due to differences in courtship and mating behaviors, as a result of which little or no sexual attraction between species, which prevents individuals from different populations from mating (Dopman et al., 2004). Similarly, when organisms are adapted to dissimilar habitats create a habitat barrier to gene flow. The different breeding times of two species can also inhibit gene flow and hence create temporal separation. Mechanical isolation occurs due to differences in size or shape of genitalia between species, which hinder the transfer of sperm, thus making interbreeding difficult or impossible, for example, in angiosperms (Grant, 1994). In the case of gametic isolation, the transferred gametes have no attraction and cannot affect fertilization, for example, in sunflowers (Rieseberg et al., 1995). If the hybrids are less attractive and cannot get mate, it will create an extrinsic barrier while if hybrid suffer reproductive problems, low gametes, etc., it will create intrinsic gene barriers.

7.3 METHODS FOR ESTIMATING GENE FLOW

The migration of individuals and genes in space has a huge impact on important evolutionary and ecological characteristics of populations (Hanski et al., 1997). The frequency of gene movement among populations assists to decide the probability of adaptation and adaptive evolution on composite landscapes. Moreover, dispersion influences the perseverance of local populations, extinction rates of species, etc. Measuring the amount of gene flow directly is very challenging and we have now developed various indirect ways to measure the same (Abdin et al., 2017). The data required for indirect estimates of gene flow are relatively easier to gather, which adds an advantage to it. At the same time, indirect methods have their disadvantages too.

7.3.1 ISLAND MODEL AND WRIGHT'S F-STATISTICS

The measure of allele association within individuals and populations is performed by Wright's F-statistics (FST) and it is the most appropriate method with a range of elucidations for studying gene flow. The main and popular one is the variance in allele frequencies among populations, σ^2_p standardized by the mean allele frequency (p) at that locus:

$$\text{FST} = \sigma^2_p / [p\,(1 - p)].$$

For the population structure, Wright (1931) put forward island, which explains the connection among many migrants a population obtains per generation and FST. Under island model,

$$\text{FST} \approx 1/\,(4Nm + 1)$$

where "N" and "m" being population size and migration rate, respectively. With the help of molecular techniques, we can detect FST value and Nm value can also be deducted using the given equation (Whitlock and Mccauley, 1999).

7.3.2 *STEPPING STONE MODEL*

In Wrights island model, the inhabitants consist of "*n*" colonies having the same size and the probability of migration (Kimura, 1953). The stepping stone model explains the allocation of allele frequencies in a population with an infinite set of demes in various locations of the space by Cartesian coordinates.

Taking the example of a one-dimensional case, *p*(*k*) being the rate of occurrence of one allele at a biallelic locus, *k* and *p* be mean allele frequency.

In each generation, the rate of occurrence of one allele at a biallelic locus is updated with the following three steps (Duforet-Frebourg and Slatkin, 2016):

A. Exchange a proportion "*mi*" of migrants with demes at a distance *i*.
B. Exchange a proportion $m\infty$ of migrants with a deme having unchanging allele frequency *p*.
C. Sample gametes of the next generation in the population.

According to this model, when migrants are exchanged simply among neighboring demes, then *mi* = 0 and *i* > 1, thus *r* can be written as an exponential function of *k* (*k* is the steps apart) as *r*(*k*) (the correlation coefficient) and is written as

$$r(k) = e^{-\sqrt{\frac{2m\infty}{m1}}k}$$

7.4 MODES OF SPECIATION

Speciation is the way how new kind of organism is created. It can happen when a group in a species divides from other members of its species and develops new and distinctive characters. The claims of special surroundings or the characteristics of members of the new group will differentiate the new species from the old one. There are many ways of speciation; the important ways are shown in Figure 7.1 and explained in the subsequent section.

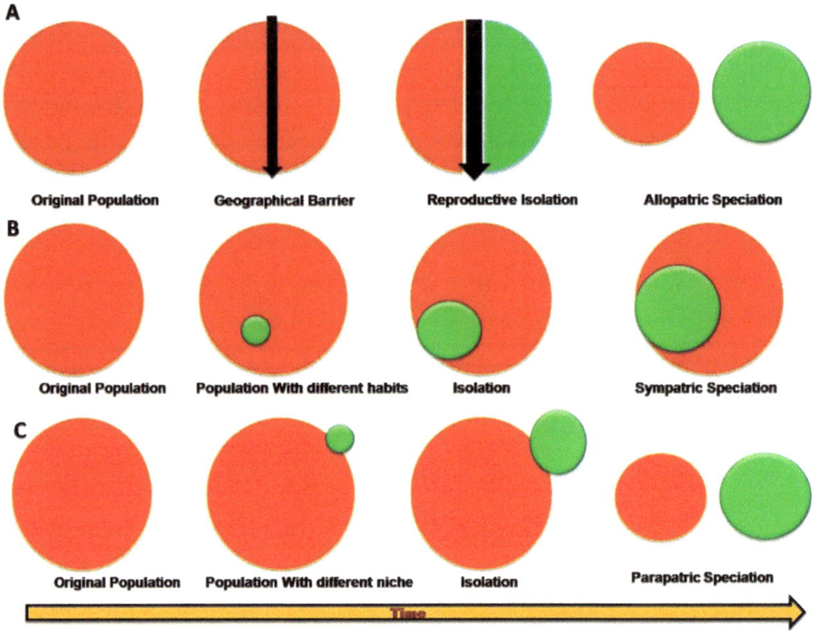

FIGURE 7.1 Modes of speciation (A) allopatric (B) sympatric (C) parapatric.

7.4.1 ALLOPATRIC SPECIATION

Allopatric speciation or geographical isolation is considered the primary mode of speciation. Birds and mammals exclusively show this type of speciation (Baptestini et al., 2013). Most evolutionary biologists believe that this mode is having an extensive role in speciation (Turelli et al., 2001). In this kind of speciation, a population gets geographically isolated from the other population of its parent species and as such forms separated subpopulations (Boucher et al., 2016, Mayr, 1963). Unsuitable areas reduce migrations. For example, a geographical barrier like a mountain range or waterway makes it impossible for the species to breed with one another. As a result, during this period of isolation, independent evolution of subspecies occurs with time, acquire characters that promote or guarantee reproductive isolation after the external barriers break down, and, thus, speciate. The spatial separation of populations, by prevention of gene flow between them, results in the accumulation of different changes in allele frequencies in those populations. These differences may lead to developmental incompatibilities, postzygotic isolation, and reproductive barriers. Such populations may not be able to mate if they come into contact again, or may form only in viable hybrids; if fertilization takes place, for example, Darwin's finches of Galapagos Islands are geographically isolated from related birds of South American mainland to form new species as a result of geographic isolation.

7.4.2 SYMPATRIC SPECIATION

Sympatric speciation, from the Greek "same place," involves the division of a parental species into two or more reproductively isolated groups without geographical isolation between these groups (Osborne et al., 2020). In this type, there are no physical barriers preventing members of a species from mating with one another, and all members live close to one another (Foote, 2018). It occurs when a subpopulation that occupies the same niche as the remainder of the species develops a unique mutation, perhaps based on a food source or characteristic that prevents it from mating with the original population. Polyploidy can cause sympatric speciation (Ramsey and Schemske, 1998). Reproductive isolation can also be caused by chromosome doubling and there is no necessity for geographical isolation. The cases of this speciation are rare but it does occur (Dowling and Secor, 1997).

7.4.3 PARAPATRIC SPECIATION

Parapatric (para, near) speciation happens when a tiny population is cut off, generally at the periphery of a huge population, and acquires variability up the mark that they speciate (Hollinger and Hermisson, 2017). The flow of genes may be possible during this process and hybrid zones are visible. The grass *Agrostis tenuis* shows this kind of speciation (McNeilly and Antonovics, 1968). Parapatric speciation happens when new species evolve in continuous, yet spatially segregated habitats. Different ecological habits, different mutations by selection or drift, and variable acclimatization can lead to parapatric speciation (Gavrilets, 2003). The flow rate of genes is lower in allopatric and more in parapatric speciation.

7.5 GENETIC DRIFT

Genetic drift is defined as the random variations in allele frequencies from generation to generation, which is not due to natural selection. It is also known as "random drift in allele frequencies." Because this theory of genetic drift was developed by geneticist Sewall Wright, it is also called the Sewall Wright effect or scattering of variability. Because of this effect, even a new or rare allele might come to possess some unusual characteristics and might dominate a population over several generations, especially in small populations (Higgins and Lynch, 2001). Due to the phenomenon of genetic drift, some alleles will eventually become fixed in a population, while alternate alleles will eliminate (Zeigler, 2014). Hence, through loss of genes or fixation of alleles in a population, drift results in a reduction in genetic variation. Genetic drift causes an increase in genetic diversity among subpopulations that have limited gene flow, with the degree of genetic variability increasing with time (Ortiz-Barrientos and James, 2017).

In allele frequency, genetic drift results in extra random alterations and more fast loss of genes or fixation of alleles in populations with smaller sizes. This genetic drift has a role in all limited populations, but in small populations, its effects are much prominent (Templeton, 2019). According to Wright (1931), significant changes will be caused by drift if 1 /4Ne values are much greater than the mutation rate and the selection coefficient (s), where Ne is the effective population size. The number of individuals in an ideal population that would have the same genetic response to random processes as a real population of size *N* depicts the effective population size (Crawford, 1984; Wright, 1931).

7.6 INBREEDING AND INBREEDING DEPRESSION

The phenomenon of "inbreeding" refers to all those situations in which mating occurs among relatives or individuals linked by common descent. Such mating is associated with increased homozygosis. If the parents of an individual were strongly related to each other than randomly chosen individuals, then the individual is said to have been "inbred." When populations are partitioned into tiny clusters, inbreeding will take place purely due to restricted population size and genetic drift. It takes place even if mating occurs as a chance in subpopulaces (Jacquard, 1975). Therefore, the extent to which specific mating is an inbred is based on ancestors. Briefly, the word inbreeding is defined as mating involving close families (first cousins or closer) (Blouin and Blouin, 1988). Inbreeding depression, the decrease in fitness (lower survival, mating, or reproduction) due to inbreeding, is a virtually widespread fact that depends on mutation, drift, and selection (Charlesworth and Willis, 2009). The degree of variation in inbreeding depression also depends upon the phylogeny of the species. Moreover, the intensity of depression for factors of fitness like survival, mating can also differ and as a rule enhances when the surrounding is under stress or competition (Armbruster and Reed, 2005; Fox and Reed, 2011). Inbreeding decreases vigor due to genetic or inbreeding load and is expressed mostly in homozygotes. Inbreeding load is an active character, which is believed to be driven by continuous mutation.

7.6.1 MEASUREMENT OF INBREEDING DEPRESSION WITHIN THE POPULATION

Normally, inbreeding depression is calculated by relating the fitness of progeny from outcrossed individuals with inbred crosses (from self-fertilization or between relatives) (Cheng et al., 2018). The reduction in fitness of selfed progeny relative to outcrossed progeny can then be described as

$$\sigma = (Wo - Ws)/Wo = 1 - Ws/W$$

where Wo is the fitness of outcrossed progeny and Ws is the fitness of selfed progeny.

7.7 GENETIC BOTTLENECK AND FOUNDER EFFECT

A "population bottleneck" or "genetic bottleneck" is a yearly or seasonal phenomenon of cyclic fluctuation in population density because of ecological events, for instance, food shortages, earth din, droughts, floods, flames, and infection, causing periodic squeezing of some of the genes in a gene pool in a random fashion. The occurrence of these events can decrease the variability in the gene pool, after that only a few individuals with a lesser genetic diversity form the progenitors for the future generation of the population, which may pass on genes to the next generations through sexual reproduction. Genetic diversity remains lower, growing only when gene flow from another population takes place or increases very slowly with time as random mutations occur. Alternatively, if the surviving population has huge genetic fitness, the presence of better genes in the gene pool of the population is greater than before whereas the pool is reduced. The drift due to population bottleneck can change the proportionate random distribution of genes and also lead to loss of alleles (Lynch et al., 1995). The strength of a bottleneck depends on two things: first, the extent to which population size is reduced, and secondly, the duration of the bottleneck: if the restriction in size is smaller and the time duration is longer, the bottleneck effect will be more severe. Both these parameters influence the genetic variation in a bottlenecked population.

Founder effect or founder principle implies the setting up of a new population by a few individuals who become founders or founder members; they have a limited number of genes than the parents (Ladizinsky, 1985). The result of this phenomenon is low genetic variability in the new population (Mayr, 1942). This effect looks like the specialized event of genetic drift, which occurs if a small population fragments off from the parent population and establishes a new one. The new colony may have a low frequency of alleles than the original population. This makes the colony more vulnerable to extinction (Reece et al., 2011). Figure 7.2 depicts the founder and bottleneck effect in the population.

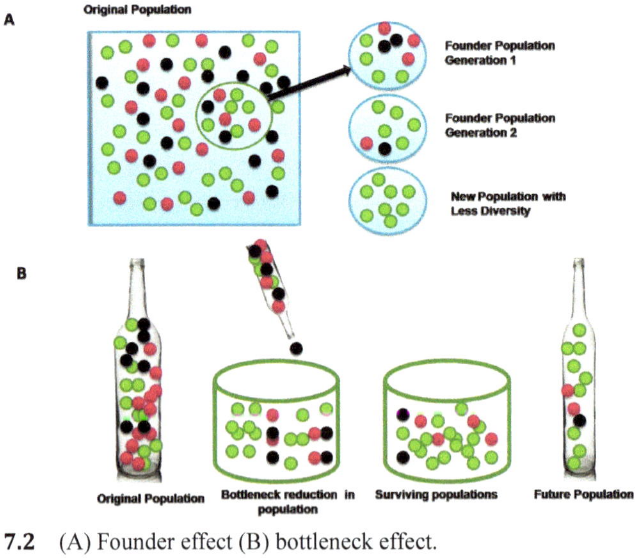

FIGURE 7.2 (A) Founder effect (B) bottleneck effect.

7.8 DISPERSAL IN PLANT POPULATION

The distribution and richness of species in an area are determined by the location of individuals and their movement between different locations. Understanding dispersal is important for foretelling the capability of the population to withstand environmental changes (Watkinson and Gill, 2002), range expansion of alien species (Higgins et al., 2001), or reestablishment by rare species (Coulson et al., 2001). In plants, propagation occurs through seeds and spores (Petit et al., 2005) but fragmentation, bulbils, or budding is also common. Many species have more than one type of method, for example, *Fallopia japonica* (Forman and Kesseli, 2003), *Undaria pinnatifida* (Forrest et al., 2000), etc. Some propagules are dispersed by flowing water currents, some by fast wind, and some by the soil when taken from one place to other (Forrest et al., 2000). Spores, wind, water, and soil are cardinal for lower plant dispersal (Sundberg and Rydin, 2000). In the case of higher plants, a wide range of dispersal strategies is found (Willson and Traveset, 2000), for example, by explosive pods, wind, water, attaching to other animal species, passed from place to place by invertebrates (Ohkawara and Higashi, 1994). Anemophily seeds may experience the second dispersal after they come to rest (Schurr et al., 2005).

7.8.1 MEASUREMENT OF DISPERSAL

To study plant distribution and species richness at various levels of biodiversity, we need to know how to measure plant dispersal. To measure dispersal is a tricky task and little is known about it. We can use both direct and indirect methods of seed dispersal to measure it (Oddou-muratorio and Klein, 2008). Some of them are explained in the subsequent sections.

7.8.2 DIRECT METHODS FOR MEASURING DISPERSAL

Direct measurements are hard to attain in natural systems (Bullock et al., 2006). By tracing individual propagules or by knowing the number of propagules that arrive at a given place, in traps known as tracking and trapping methods, respectively, we can measure dispersal distances directly.

As per Nathan et al. (2003), tracking techniques are superior as compared to trapping as it provides precise dispersal distances also allows us to measure long dispersals while trapping methods needs suppositions related where propagules will drop and depend on boundaries of trapping field. Technically, in the case of animal-dispersed propagules, we can use the tracking method because animals can be tracked with convenience. As opposed to following animals, we can use an easy way of gathering propagules at a given location with the help of a trap. The achievement of such a technique is influenced by the area and site of traps. A realistic disadvantage of this method is contamination possibilities in traps with propagules from co-specifics excluding the planned ones (Uriarte et al., 2005).

Seed tracking generally involves tagging and the procedure of tagging should fulfill some necessities like it must not alter the properties, should permit easy recapture over larger areas, and should continue over the required study period. Color markings (Aerts et al., 2006), marking by thread or wire (Xiao et al., 2006), radioisotope labeling (Vander Wall et al., 2006), use of radio transmitters (Pons and Pausas, 2007), and some magnet-based locators are cardinal methods (Alverson and Diaz, 1989). Nowadays, we use fluorescent marking, ultraviolet (UV) lights, and lightweight flashlights with UV light-emitting diodes in areas with light ground vegetation because of high-applicability recapture rates (Reiter et al., 2006). This technique suits both small and large diaspores (Longland and Clements, 1995).

7.8.3 MEASURING DISPERSAL BY INDIRECT TECHNIQUES

Genetic methods are good ways to study dispersal. The use of genetic markers for measuring dispersal has been proposed by various studies (Cain et al., 2000; Nathan et al., 2003). ISSR, RAPD, VNTRs, RFLPs, start codon-targeted (SCoT) polymorphism (Collard and Mackill, 2009), and DNA fingerprints can likely provide means for direct observation (Abdin et al., 2017). In studies that have taken place over some years and genetic markers have been used to characterize population, it is likely to identify dispersal of private or rare propagules from one population to another. With the great advancements in the development of mechanistic models', movement of propagules in the air can be detected (Soons et al., 2004). We can also monitor or radio-track vertebrates (Holbrook and Smith, 2000).

7.9 DISPERSAL, GENE FLOW, STRUCTURE OF POPULATION, AND MOLECULAR MARKERS

In nature, long-space dispersal measures are extremely significant and molecular markers are likewise cardinal for studying dispersal (Van de Paer et al., 2018). Molecular markers may be either dominant or co-dominant. The application of markers has resulted in the development of techniques that help in accurately calculating gene flow. Microsatellites have facilitated the calculation of gene flow based on parental analyses. ISSR, RAPD, RFLP, AFLP, minisatellite and microsatellites, and SSR are the most widely used molecular markers (Abdin et al., 2017; Ouborg et al., 1999). RFLP technique is used to analyze and gather data on mutations taking place at particular regions called restriction sites (Botstein et al., 1980). It helps in detecting polymorphisms at the DNA level and the construction of linkage maps. In RFLP, an organism is distinguished based on examination of patterns obtained by their DNA. Organisms vary in cleavage sites and the length of pieces created will differ after DNA is digested with the help of a restriction enzyme. The species can be differentiated based on pattern similarity. In the case of gene flow and dispersal studies, nuclear RFLPs are within a given species. Certainly, genetic diversity is lesser in crop population as compared with wild populations (Marwal et al., 2014). RAPD is a PCR-dependent molecular method that confirms the diversity in the bands amplified in individuals with genetic distinction. The quantity of amplified fragments obtained is based on the allotment and number of annealing sites all through the genome. Polymerase chain reaction results are analyzed on agarose gel. With respect to markers, RAPD offers detection of a limitless quantity of loci. Taxon identification, reproductive behavior, hybridization, population genetic structure, genes of importance, inbreeding, etc., have been widely studied by this marker. RAPD plays a vital role if the pedigree of an individual is not known.

AFLP allows the real-time study of various DNA regions, which are arbitrarily spread in the genome. This dependent on selective amplification of a subset of genomic parts obtained with restriction enzymes digestion. In plants, animals, and microbes, AFLP markers are widely used in studying population diversity (Savelkoul et al., 1999). Booming claims of AFLP in environmental and evolutionary studies are documented (Bonin et al., 2007). Polymorphic AFLP loci are used for the absence and presence of heterozygous, bands, etc. Diversity arrays technology (DArT) is a technique for studying thousands of loci in one assay (Jaccoud et al., 2001). It produces full genome fingerprints for the absence and presence of DNA fragments. It can be used for the fast characterization of hundreds to thousands of polymorphisms at an affordable cost. For genetic diversity, gene flow studies, and genetic mapping of associated studies, this method has been widely used in both non-model and model crops. SCoT markers are based on the short-conserved region in plant genes. These markers are used to assess structure, genetic diversity, identify cultivar identification, QTL mapping, DNA fingerprinting in various species, such as sugarcane, rice, etc. (Collard and Mackill, 2009).

7.10 CONCLUSION

The measurement of dispersal, gene flow, and population structure are cardinal. To a large extent, direct assessment of dispersal has improved. Radio transmitters and marker-assisted approximations have allowed observation of dispersal events precisely but these methods are labor-intensive. FST is a way of measuring population dispersals and certainly knowing about the genetic structure, evolutionary properties, etc. Pieces of evidence point out that inbreeding causes an extreme decline in fitness and populations with less size are likely to require genetic management to avoid extinction. One opportunity to reduce inbreeding depression and reinstate sustainable population size is a genetic rescue, but, at the same time, habitat should be restored. Alternatively, ex-situ conservation, breeding strategies, etc. should be followed for minimizing inbreeding depression and genetic loss.

KEYWORDS

- **speciation**
- **inbreeding**
- **dispersal**
- **island model**
- **Wright's F statistics**
- **Sewall Wright effect**

REFERENCES

Abdin, M. Z.; Arya, L.; Verma, M. Use of SCoT Markers to Assess the Gene Flow and Population Structure among Two Different Populations of Bottle Gourd. *Plant Gene* **2017,** *9*, 80–86.

Aerts, R.; Maes, W.; November, E.; Behailu, M.; Poesen, J.; Deckers, J.; Muys, B. Surface Runoff and Seed Trapping Efficiency of Shrubs in a Regenerating Semiarid Woodland in Northern Ethiopia. *Catena* **2006,** *65* (1), 61–70.

Alverson, W. S.; Diaz, A. G. Measurement of the Dispersal of Large Seeds and Fruits with a Magnetic Locator. *Biotropica* **1989,** *21*, 61–63.

April, J.; Mayden, R. L.; Hanner, R. H.; Bernatchez, L. Genetic Calibration of Species Diversity among North America's Freshwater Fishes. *Proc. Natl. Acad. Sci.* **2011,** *108* (26), 10602–10607.

Armbruster, P.; Reed, D. H. Inbreeding Depression in Benign and Stressful Environments. *Heredity* **2005,** *95* (3), 235–242.

Baptestini, E. M.; de Aguiar, M. A.; Bar-Yam, Y. Conditions for Neutral Speciation via Isolation by Distance. *J. Theoretical Biology* **2013,** *335*, 51–56.

Blouin, S. F.; Blouin, M. Inbreeding Avoidance Behaviors. *Trends Ecol. Evol.* **1988,** *3* (9), 230–233.

Bolstad, G. H.; Hindar, K.; Robertsen, G.; Jonsson, B.; Sægrov, H.; Diserud, O. H.; Barlaup, B. T. Gene Flow from Domesticated Escapes Alters the Life History of Wild Atlantic Salmon. *Nat. Ecol. Evol.* **2017,** *1* (5), 1–5.

Bonin, A.; Ehrich, D.; Manel, S. Statistical Analysis of Amplified Fragment Length Polymorphism Data: A Toolbox for Molecular Ecologists and Evolutionists. *Mol. Ecol.* **2007,** *16* (18), 3737–3758.

Botstein, D.; White, R. L.; Skolnick, M.; Davis, R. W. Construction of a Genetic Linkage Map in Man Using Restriction Fragment Length Polymorphisms. *Am. J. Hum. Genet.* **1980,** *32* (3), 314.

Boucher, F. C.; Zimmermann, N. E.; Conti, E. Allopatric Speciation with Little Niche Divergence Is Common among Alpine Primulaceae. *J. Biogeogr.* **2016,** *43* (3), 591–602.

Bullock, J. M.; Shea, K.; Skarpaas, O. Measuring Plant Dispersal: An Introduction to Field Methods and Experimental Design. *Plant Ecol.* **2006,** *186* (2), 217–234.

Cain, M. L.; Milligan, B. G.; Strand, A. E. Long-Distance Seed Dispersal in Plant Populations. *Am. J. Bot.* **2000,** *87* (9), 1217–1227.

Charlesworth, D.; Willis, J. H. The Genetics of Inbreeding Depression. *Nat. Rev. Genet.* **2009,** *10* (11), 783–796.

Cheng, F.; Wu, J.; Cai, X.; Liang, J.; Freeling, M.; Wang, X. Gene Retention, Fractionation and Subgenome Differences in Polyploid Plants. *Nat. Plants* **2018,** *4* (5), 258–268.

Collard, B. C.; Mackill, D. J. Start Codon Targeted (SCoT) Polymorphism: A Simple, Novel DNA Marker Technique for Generating Gene-Targeted Markers in Plants. *Plant Mol. Biol. Rep.* **2009,** *27* (1), 86.

Coulson, S. J.; Bullock, J. M.; Stevenson, M. J.; Pywell, R. F. Colonization of Grassland by Sown Species: Dispersal versus Microsite Limitation in Responses to Management. *J. Appl. Ecol.* **2001,** *38* (1), 204–216.

Crawford, T. J. What Is a Population? In *Evolutionary Ecology*; Shorrocks, B., Ed.; Oxford: Blackwell Scientific, 1984; pp 135–174.

Donnelly, M. J.; Townson, H. Evidence for Extensive Genetic Differentiation among Populations of the Malaria Vector *Anopheles arabiensis* in Eastern Africa. *Insect Mol. Biol.* **2000,** *9* (4), 357–367.

Dopman, E. B.; Bogdanowicz, S. M.; Harrison, R. G. Genetic Mapping of Sexual Isolation between E and Z Pheromone Strains of the European Corn Borer (*Ostrinia nubilalis*). *Genetics* **2004,** *167* (1), 301–309.

Dowling, T. E.; Secor, C. L. The Role of Hybridization and Introgression in the Diversification of Animals. *Annu. Rev. Ecol. Syst.* **1997,** *28* (1), 593–619.

Duforet-Frebourg, N.; Slatkin, M. Isolation-by-Distance-and-Time in a Stepping-Stone Model. *Theor. Population Biol.* **2016,** *108*, 24–35.

Foote, A. D. Sympatric Speciation in the Genomic Era. *Trends Ecol. Evol.* **2018,** *33* (2), 85–95.

Forman, J.; Kesseli, R. V. Sexual Reproduction in the Invasive Species *Fallopia japonica* (Polygonaceae). *Am. J. Bot.* **2003,** *90* (4), 586–592.

Forrest, B. M.; Brown, S. N.; Taylor, M. D.; Hurd, C. L.; Hay, C. H. The Role of Natural Dispersal Mechanisms in the Spread of *Undaria pinnatifida* (Laminariales, Phaeophyceae). *Phycologia* **2000,** *39* (6), 547–553.

Fox, C. W.; Reed, D. H. Inbreeding Depression Increases with Environmental Stress: An Experimental Study and Meta-Analysis. *Evol.: Int. J. Org. Evol.* **2011,** 65 (1), 246–258.

Garcia, C.; Jordano, P.; Godoy, J. A. Contemporary Pollen and Seed Dispersal in a *Prunus mahaleb* Population: Patterns in Distance and Direction. *Mol. Ecol.* **2007,** *16* (9), 1947–1955.

Gavrilets, S. Perspective: Models of Speciation: What Have We Learned in 40 Years? *Evolution* **2003,** *57* (10), 2197–2215.

Grant, V. Modes and Origins of Mechanical and Ethological Isolation in Angiosperms. *Proc. Natl. Acad. Sci.* **1994,** *91* (1), 3–10.

Hanski, I. A.; Gilpin, M. E.; McCauley, D. E. Metapopulation Biology. In Hanski, I., Gilpin, M. E., Eds.; Academic Press: San Diego, CA, 1997; pp 454, 1–512.

Higgins, K.; Lynch, M. Metapopulation Extinction Caused by Mutation Accumulation. *Proc. Natl. Acad. Sci.* **2001,** *98* (5), 2928–2933.

Higgins, S. I.; Richardson, D. M.; Cowling, R. M. Validation of a Spatial Simulation Model of a Spreading Alien Plant Population. *J. Appl. Ecol.* **2001,** *38* (3), 571–584.

Holbrook, K. M.; Smith, T. B. Seed Dispersal and Movement Patterns in Two Species of *Ceratogymna hornbills* in a West African Tropical Lowland Forest. *Oecologia* **2000,** *125* (2), 249–257.

Hollinger, I.; Hermisson, J. Bounds to Parapatric Speciation: A Dobzhansky–Muller Incompatibility Model Involving Autosomes, X Chromosomes, and Mitochondria. *Evol.* **2017,** *71* (5), 1366–1380.

Imbert, E.; Lefèvre, F. Dispersal and Gene Flow of *Populus nigra* (Salicaceae) along a Dynamic River System. *J. Ecol.* **2003,** *91* (3), 447–456.

Jaccoud, D.; Peng, K.; Feinstein, D.; Kilian, A. Diversity Arrays: A Solid State Technology for Sequence Information Independent Genotyping. *Nucl. Acids Res.* **2001,** *29* (4), e25–e25.

Jacquard, A. Inbreeding: One Word, Several Meanings. *Theor. Population Biol.* **1978,** *7* (3), 338–363.

Kimura, M. Stepping Stone Model of Population. *Annu. Rep. Natl. Inst. Genet. Jpn* **1953,** *3*, 62–63.

Ladizinsky, G. Founder Effect in Crop-Plant Evolution. *Econ. Bot.* **1985,** *39* (2), 191–199.

Longland, W. S.; Clements, C. Use of Fluorescent Pigments in Studies of Seed Caching by Rodents. *J. Mammal.* **1995,** *76* (4), 1260–1266.

Lynch, M.; Conery, J.; Burger, R. Mutation Accumulation and the Extinction of Small Populations. *Am. Natural.* **1995,** *146* (4), 489–518.

Marwal, A.; Sahu, A. K.; Gaur, R. K. Molecular Markers: Tool for Genetic Analysis. In *Animal Biotechnology*; Verma, A. S., Singh, A., Eds.; Academic Press, 2014; pp 289–305. ISBN 9780124160026.

Mayr, E. *Animal Species and Evolution*; Belknap Press of Harvard University Press: Cambridge, MA, 1963; p 604.

Mayr, E. *Systematics and the Origin of Species*; Columbia University Press: New York, 1942.

McNeilly, T. Evolution in Closely Adjacent Plant Populations III. Agrostis Tenuis on a Small Copper Mine. *Heredity* **1968,** *23* (1), 99–108.

Nathan, R.; Perry, G.; Cronin, J. T.; Strand, A. E.; Cain, M. L. Methods for Estimating Long-Distance Dispersal. *Oikos* **2003,** *103* (2), 261–273.

Nosil, P. *Ecological Speciation*; Oxford University Press: New York, 2012.

Oddou-muratorio, Sylvie.; Klein, E. K. Comparing Direct vs. Indirect Estimates of Gene Flow within a Population of a Scattered Tree Species. *Mol. Ecol.* **2008,** *17* (11), 2743–2754.

Ohkawara, K.; Higashi, S. Relative Importance of Ballistic and Ant Dispersal in Two Diplochorous *Viola* Species (Violaceae). *Oecologia* **1994,** *100* (1), 135–140.

Ortiz-Barrientos, D.; James, M. E. Evolution of Recombination Rates and the Genomic Landscape of Speciation. *J. Evol. Biol.* **2017,** *30* (8), 1519–1521.

Osborne, O. G.; Kafle, T.; Brewer, T.; Dobreva, M. P.; Hutton, I.; Savolainen, V. Sympatric Speciation in Mountain Roses (Metrosideros) on an Oceanic Island. *Phil. Trans. R. Soc. B* **2020,** *375* (1806), 20190542.

Ouborg, N. J.; Piquot, Y.; Van Groenendael, J. M. Population Genetics, Molecular Markers and the Study of Dispersal in Plants. *J. Ecol.* **1999,** *87* (4), 551–568.

Petit, R. J.; Duminil, J.; Fineschi, S.; Hampe, A.; Salvini, D.; Vendramin, G. G. Invited Review: Comparative Organization of Chloroplast, Mitochondrial and Nuclear Diversity in Plant Populations. *Mol. Ecol.* **2005,** *14* (3), 689–701.

Pons, J.; Pausas, J. G. Acorn Dispersal Estimated by Radio-Tracking. *Oecologia* **2007,** *153* (4), 903–911.

Ramsey, J.; Bradshaw Jr, H. D.; Schemske, D. W. Components of Reproductive Isolation between the Monkey Flowers *Mimulus lewisii* and *M. cardinalis* (Phrymaceae). *Evolution* **2003,** *57* (7), 1520–1534.

Ramsey, J.; Schemske, D. W. Pathways, Mechanisms, and Rates of polyploid Formation in Flowering Plants. *Annu. Rev. Ecol. Syst.* **1998,** *29* (1), 467–501.

Reece, J. B.; Urry, L. A.; Cain, M. L.; Wasserman, S. A.; Minorsky, P. V.; Jackson, R. B. AP Edition Campbell Biology, 2011.

Reiter, J.; Curio, E.; Tacud, B.; Urbina, H.; Geronimo, F. Tracking Bat-Dispersed Seeds Using Fluorescent Pigment 1. *Biotropica: J. Biol. Conserv.* **2006,** *38* (1), 64–68.

Rieseberg, L. H.; Van Fossen, C.; Desrochers, A. M. Hybrid Speciation Accompanied by Genomic Reorganization in Wild Sunflowers. *Nature* **1995,** *375* (6529), 313–316.

Rousset, F. Genetic Differentiation between Individuals. *J. Evol. Biol.* **2000,** *13*, 58–62.

Savelkoul, P. H. M.; Aarts, H. J. M.; De Haas, J.; Dijkshoorn, L.; Duim, B.; Otsen, M.; Lenstra, J. A. Amplified-Fragment Length Polymorphism Analysis: The State of an Art. *J. Clin. Microbiol.* **1999,** *37* (10), 3083–3091.

Schurr, F. M.; Bond, W. J.; Midgley, G. F.; Higgins, S. I. A Mechanistic Model for Secondary Seed Dispersal by Wind and Its Experimental Validation. *J. Ecol.* **2005,** *93* (5), 1017–1028.

Soons, M. B.; Heil, G. W.; Nathan, R.; Katul, G. G. Determinants of Long-Distance Seed Dispersal by Wind in Grasslands. *Ecology* **2004,** *85* (11), 3056–3068.

Sundberg, S.; Rydin, H. Experimental Evidence for a Persistent Spore Bank in Sphagnum. *New Phytol.* **2000,** *148* (1), 105–116.

Templeton, A. R. Genetic Drift. In *Human Population Genetics and Genomics*; Templeton, A. R., Ed.; Academic Press, 2019; pp 101–127, ISBN 978012386**0255,** https://doi.org/10.1016/B978-0-12-386025-5.00004-X

Turelli, M.; Barton, N. H.; Coyne, J. A. Theory and Speciation. *Trends Ecol. Evol.* **2001,** *16* (7), 330–343.

Turner, M. E.; Stephens, J. C.; Anderson, W. W. Homozygosity and Patch Structure in Plant Populations as a Result of Nearest-Neighbor Pollination. *Proc. Natl. Acad. Sci.* **1982,** *79* (1), 203–207.

Uriarte, M.; Canham, C. D.; Thompson, J.; Zimmerman, J. K.; Brokaw, N. Seedling Recruitment in a Hurricane-Driven Tropical Forest: Light Limitation, Density-Dependence and the Spatial Distribution of Parent Trees. *J. Ecol.* **2005,** *93* (2), 291–304.

Van de Paer, C.; Bouchez, O.; Besnard, G. Prospects on the Evolutionary Mitogenomics of Plants: A Case Study on the Olive Family (Oleaceae). *Mol. Ecol. Resour.* **2018,** *18* (3), 407–423.

Vander Wall, S. B.; Borchert, M. I.; Gworek, J. R. Secondary Dispersal of Bigcone Douglas-fir (*Pseudotsuga macrocarpa*) Seeds. *Acta Oecologica* **2006,** *30* (1), 100–106.

Wang, H.; Hou, L.; Liu, Y.; Liu, K.; Zhang, L.; Huang, F.; Yu, C. Horizontal and Vertical Gene Transfer Drive Sediment Antibiotic Resistome in an Urban Lagoon System. *J. Environ. Sci.* **2021,** *102*, 11–23.

Watkinson, A. R.; Gill, J. A. Climate Change and Dispersal. In *Dispersal Ecology*; Bullock, J. M., Kenward, R. E., Hails R. S., Eds.; Blackwell Publishing: Oxford, 2001; pp 410–428.

Whitlock, M. C.; Mccauley, D. E. Indirect Measures of Gene Flow and Migration: FST≠ 1/(4Nm+1). *Heredity* **1999,** *82* (2), 117–125.

Willson, M. F.; Traveset, A. The Ecology of Seed Dispersal. In *Seeds: The Ecology of Regeneration in Plant Communities*; Fenner, M., Ed., Vol. 2; CABI: Wallingford, 2000; pp 85–110.

Wright, S. Evolution in Mendelian Populations. *Genetics* **1931,** *16* (2), 97.

Xiao, Z.; Jansen, P. A.; Zhang, Z. Using Seed-Tagging Methods for Assessing Post-Dispersal Seed Fate in Rodent-Dispersed Trees. *Forest Ecol. Manage.* **2006,** *223* (1–3), 18–23.

Zeigler, D.; Genetic Drift. *Evolution*; Zeigler, D., Ed.; Academic Press, 2014; pp 79–84. ISBN 9780128003480. https://doi.org/10.1016/B978-0-12-800348-0.00009-2

CHAPTER 8

Ecological Genomics in Context with Plant Adaptations and Speciation

SHEIKH TAJAMUL ISLAM[1], BASHARAT AHMAD BHAT[1], INAYATULLAH TAHIR[1], and TANVIR UL HASSAN DAR[2,*]

[1]Department of Bioresources, School of Biological Sciences, University of Kashmir, Srinagar, Jammu and Kashmir, India

[2]Department of Biotechnology, School of Biosciences and Biotechnology, BGSB University, Rajouri, Jammu and Kashmir, India

[*]Corresponding author. E-mail: tanvirulhasan@gmail.com

ABSTRACT

The ecological genomics approach aims to unravel the genetic basis underlying responses of organisms to the environment. It helps to understand the structural and functional aspects of genes with ecological and evolutionary significance. The survival chances of a species in changing environment depend upon how it responds to the modified environment. It involves adaptation response expressed by an organism at morphological, anatomical, physiological, or reproductive levels to cope with the environment. Using genomic tools, it is possible to unravel phenotypic variation's genetic basis and understand how organisms adapt to the climatic variation that ultimately leads to reproductive isolation and speciation. Genomic approaches, such as quantitative trait locus studies, genome-wide association studies, genome scans, and studies on inversions allow for a better understanding of adaptation and validating various theories concerning

Plant Ecogenomics. Peerzada Arshid Shabir, PhD, Peerzada Yasir Yousuf, PhD & Khalid Rehman Hakeem, PhD (Eds.)

speciation. By these techniques, it is possible to identify and map ecologically and evolutionarily relevant genes and genomic regions based on their linkage to identifiable markers. Whole-genome polymorphism data from large populations can be used to detect genes related to adaptive phenotypic variations and to scan the footprints of selection across the genome. Transcriptome and proteome help evaluate gene expression changes, separate and identify the proteins to observe the difference in their expression in response to variation in environmental parameters or a specific ecological interaction. Genomic studies on speciation and adaptation in plants specifically focus on adaptation to biotic and abiotic environments, particularly local adaptation, reproductive isolation, gene flow, and genetic basis of abrupt speciation through interspecific hybridization.

8.1 INTRODUCTION

Ecological genomics integrates genomic and ecological approaches to unravel the genetic basis underlying biochemical, physiological, morphological, or behavioral responses of organisms to the environment. Genomic approaches include any genome-enabled process aimed to understand the structural aspects or ecological functions of genes (Ungerer et al., 2008). The role of genomics in ecology can be understood from how it has helped to understand the biological phenomena, such as allelopathy and plant–microbe interactions that have remained a focus in ecology (Baldwin, 2003; Ungerer et al., 2008). For example, gene expression studies in susceptible *Arabidopsis thaliana*, treated with (-)-catechin from the roots of *Centaurea maculosa* (spotted knapweed) confirmed allelopathy as a possible mechanism used by spotted knapweed to out-compete native plants in parts of North America (Bais et al., 2003).

Species are considered a set of closely resembling individuals that can interbreed to produce fertile offsprings (Mayr, 1942). Speciation is an exciting phenomenon that has fascinated evolutionary biologists. New species originate either by the process of ecological speciation or by mutation-order speciation, and it is the natural selection that drives the fixation of alleles in both cases (Schluter and Conte, 2009). The natural environment is not constant, and it presents a combination of biotic and abiotic challenges which require ecological and evolutionary responses from a species for adaptation and survival. Each population of a widespread species has relatively higher fitness to its native site than the population introduced from other sites, a result of environment–genome interactions for Darwinian fitness. In general, adaptation refers to phenotypic variation, which improves reproductive success and chances of survival relative to alternative features (Gregory, 2009).

Through genomic tools, it has become possible to unravel phenotypic variation's genetic basis and understand how adaptation to local climate ultimately leads to speciation. Genomics can help identify genomic regions and metabolic networks involved in the adaptive response to biotic and abiotic environments (Haak et al., 2014).

8.2 SPECIES AND SPECIATION

Species are usually recognized as closely resembling individuals that can interbreed freely to produce fertile offsprings, conferring importance to reproductive isolation. Species being the most critical unit in biology, and speciation is the fundamental process contributing to existing biodiversity (Nosil and Feder, 2012). In ecological speciation, divergent selection drives reproductive isolation between populations, fixing different advantageous alleles in different environments. In mutation-order speciation, diverse populations accumulate other series of incompatible mutations and diverge as they adapt to similar selection regimes (Schluter and Conte, 2009). Although various models explain speciation, we still lack a clear understanding of speciation presumably because it is usually a slow process, too long to observe directly. It is complex involving multiple factors acting in favor or against it through time and space (Nevo, 2006). Populations are subjected to varied environmental and demographic processes. The novel ecological opportunities that arise resulting from it provide opportunities for divergence where selection can act on variation to shape evolution. Although multiple factors are operative at any given time, genetic loci that reduce gene flow or reduce hybrid fitness between diverging genomes may accumulate over time and favor reproductive isolation, a key factor for speciation. It has been suggested that speciation can occur even without geographic separation. It depends on the strength of disruptive selection and linkage relationships of genes under disruptive selection (Abbott et al., 2013; Maan and Seehausen, 2011).

Genomic studies on speciation usually focus on variation, adaptation to local climate, reproductive isolation, and gene flow. Hybridization is being looked at as an essential tool for rapid speciation. Ecological genomics helps genotype–phenotype mapping that connects patterns of phenotypic variation to genetic mechanisms. It facilitates understanding the evolution of trait variation under selection pressure and isolation mechanisms (e.g., reproductive isolation) both of which are essential elements for speciation (Haak et al., 2014; Pinho and Hey, 2010).

8.3 ADAPTATION

Environmental conditions vary along latitudes, longitudes, and altitudes. These variations are usually gradual; however, a sharp change in the plant environment can be observed over a short distance, for example, change related to soil type, moisture content, slope, temperature, and light incidence. The survival chances of a species in a changing environment depend upon how it responds to the modified environment. It involves adaptation response expressed by an organism at morphological, anatomical, physiological, or reproductive levels to cope with the environment. Adaptation affects the abiotic environment's response and includes the biotic environment's response, for example, response during plant–microbe interaction and herbivory (Abbott, 2017; Bamba et al., 2019). Adaptation may come as a rapid response to changing the environment in terms of phenotypic plasticity or may come through adaptive genetic divergence. At times, phenotypic plasticity can shield genes from natural selection and retard adaptation, as it can shift the distribution of phenotypes in the population closer to the optimum. Plasticity is considered the primary mechanism in explaining phenotypic change under environmental gradient, except if direct evidence for the genetic basis of the variation is available. However, plant species response to environmental gradients through adaptive genetic changes may provide initial input to form fully reproductively isolated forms or biological species (Abbott, 2017; Fox et al., 2019).

8.4 GENOMICS IN ADAPTATION AND SPECIATION

The availability of genomic tools has opened up an opportunity to validate various theories concerning speciation and understand how reproductive isolation leads to the formation of new species (Hopkins et al., 2020). Genomic approaches such as quantitative trait locus (QTL) studies and genome-wide association studies (GWAS) aim to unravel the genetic mechanisms underlying organism phenotypic variation in response to the environment, essential to understanding how organisms can produce adaptive phenotypes (Tarazona et al., 2019). In this context, it is likely to identify the genes involved in the organism–environment interaction and investigate these genes' role in the ecologically relevant process or phenotypic expression. QTL mapping can help map ecologically and evolutionarily relevant genes based on their linkage to identifiable markers in mapping populations. Fine-scale mapping is possible if detailed recombination maps and physical maps are available (Ungerer et al., 2008; Voelckel et al., 2017). Population genomics utilizing whole-genome polymorphism data from large populations can detect genes related to adaptive phenotypic variations and scan the footprints of selection across the genome. Quantitative genetic approaches, however, are not always possible to apply especially for plant species with long generation times, polyploid genomes, and those difficult to cultivate (Voelckel et al., 2017).

Transcriptome, consisting of all RNA molecules existing in a cell or tissue, is the first genome expression product. The transcriptome is not constant and varies with stage of development, physiological, and environmental conditions. Changes in transcriptome or gene expression are analyzed using various techniques, such as quantitative reverse transcription PCR (RT-qPCR), microarrays, or sequencing (Zapalska-Sozoniuk et al., 2019). It has been suggested that to evaluate species' adaptive capacity, it is not necessary to understand all of the components of genetic variation. Sometimes expression differences of individual genes are linked to the rapid evolution of traits (Kapun et al., 2016). Microarrays help identification of genes and transcriptional analysis of gene expression changes under changing environmental parameters. Proteomic methods can be directly used to separate and identify the proteins from the samples to observe the difference in their expression in response to an environmental perturbance or a specific ecological interaction. However, in both these approaches, the candidate genes or proteins' functional relevance needs to be confirmed through further investigations (e.g., by using mutants) (Ungerer et al., 2008).

Chromosomal inversions suppress recombination and produce new mutations at their breakpoints, providing a source of genetic variation. For these reasons, chromosomal inversions have been repeatedly linked to critical evolutionary processes, such as alternate reproductive strategies, reproductive isolation, assertive mating, adaptive divergence, and speciation (Huang and Rieseberg, 2020). CRISPR/Cas9 system is a gene-editing bacterial origin technique that uses Cas9- an RNA-guided endonuclease that cleaves DNA. It is used for gene editing, targeted insertion and deletion of nucleotide, gene replacement, and structural change. It can be used to create targeted inversions or to reverse an existing inversion. Recombinants with a noninverted variant can be used to fine-mapping traits and their causal loci across the inversion (Hopkins et al., 2020).

Genomic studies in plants in ecological and evolutionary sense focus on adaptation to the biotic environment, adaptation to abiotic environments, particularly local adaptation, reproductive isolation, and gene flow. Many studies aim to understand the genetic basis of abrupt speciation through interspecific hybridization in plants (Bamba et al., 2019; Gompert et al., 2017).

8.5 ADAPTATION TO ABIOTIC ENVIRONMENT

Plants are sessile and unable to escape their environment. As such responses to variation in the environmental factors are of great importance in plants in particular. Gene expression variation is closely linked to plant adaptation. Xu et al. (2016) in a transcriptomic study on *Miscanthus lutarioriparius* populations assessed the relation between changes in gene expression concerning the difference in environment variables. Using RNA sequence data of individuals from 14 diverse populations, genetic diversity and comparison of the gene expression in native and new harsh environment revealed that gene expression variation is mainly influenced by environment and genetic diversity in changing environment. The study concluded that genes with higher expression are under strong purifying selection compared with genes with a lower level of expression. Proteome comparisons in evolutionary studies focusing on adaptation and speciation have gained importance because of the apparent close relationship between protein levels and phenotypes compared with genome and transcriptome (Manzoni et al., 2018). Selected reaction monitoring (SRM) studies or targeted proteomics mainly study-specific proteins related to a biomarker, a biochemical pathway, or a protein network. This makes SRM an attractive technique for biomarker verification and validation or monitoring pathways across multiple environments and phenotypes, particularly in studying adaptation (Drabovich et al., 2015; Ghanemi, 2014). A study on morphological, biochemical, and proteomics parameters conducted on the adaptation of *Potentilla saundersiana* to altitudinal gradient concluded that plant species employ a multipronged strategy for adaptation to high-altitude environments. Proteomics results revealed differential protein expression corresponding to the altitudinal gradient. It confirmed that proteins involved in energy transduction, scavenging of reactive oxygen species, defense process, DNA stability, and protein turnover are essential for the adaptation to environmental variation and stress (Ma et al., 2015).

The role of inversions in environmental adaptation has been suggested by various studies on patterns of clinal variation. An inversion on chromosome 1 in maize (*Zea mays*) reflects strong altitudinal influence. It shows an association with environmental variables and phenotypic traits (Fang et al., 2012). Lowry and Willis (2010) through reciprocal transplant experiments demonstrated that alternative chromosomal inversions associated with flowering time contribute to adaptation to the local environment, life history transitions, and reproductive isolation in yellow monkey flower (*Mimulus guttatus*).

8.6 LOCAL ADAPTATION

The natural environment varies in space. Each population of a widespread species has relatively higher fitness to its native site compared with the population introduced from other sites, which are referred to as local adaptation. Local adaptation represents variations in the patterns and processes observed in populations of a species occupying different habitats, due to environment–genome interactions. It can gradually lead to reproductive isolation and speciation. Genetically based phenotypic variations observed along latitudinal and altitudinal gradients provide clues for local adaptation (Zhang et al., 2019).

8.7 GENOMICS OF LOCAL ADAPTATION

Although the genetic complexities underlying local adaptation are not well understood, genes for several adaptations which show Mendelian inheritance have been characterized. Plants variously respond to abiotic and biotic stress. Genomic regions for salinity, drought stress response, and trichome variation (a defense to herbivory) have been mapped in *Arabidopsis* species. The *Arabidopsis* species have evolved multiple strategies as a defense against drought. These strategic differences align with the distinct ecological priorities of the stress-tolerant *A. lyrata*, the competitive *A. halleri* and the ruderal *A. thaliana*. Trichome variation in *Brassicaceae* is a fine example of a single plant gene's community-wide effect on herbivory (Bouzid et al., 2019; Sato et al., 2019). Genome-wide studies combined with field experiments can help unravel the genetic basis of local adaptation both in the model and nonmodel species, which can ascertain how the patterns of adaptation in nonmodel species are different from those of model species (Savolainen et al., 2013; Yoder et al., 2014). Yoder et al. (2014) obtained data of single nucleotide polymorphisms (SNP) and identified loci associated with three important climatic variables, such as annual mean temperature, wettest month precipitation, and isothermality. The identified loci have high sequence similarity with genes already known to be responsible for tolerance to temperature and drought or herbivory resistance.

Selection scans and GWAS are beneficial when studying perennial species, especially when classical crossing-based approaches are usually impossible to conduct. However, GWAS (McKown et al., 2018) identified candidate loci associated with bud-break, a crucial adaptive feature in temperate perennial woody plants. Studies on local adaptation usually compare populations separated by huge distances, preventing gene exchange, and not many studies on microgeographic divergence. However, a survey on small-scale local adaptation in *Arabidopsis halleri* subsp. *gemmifera*, involving genome-wide screening along an altitudinal gradient on two mountains in Japan identified genes responsible for trichome density and freezing tolerance related to the known phenotypic and environmental differences along altitudes (Kubota et al., 2015).

8.8 REPRODUCTIVE ISOLATION AND GENE FLOW

Reproductive isolation is considered essential for maintaining species identity. It refers to genetically based barriers that prevent gene flow and favor adaptive divergence between populations. Adaptive variation and disruptive selection under varied environmental pressures can lead to speciation (Bamba et al., 2019). Attempts to ascertain the genetic basis of reproductive isolation are essential aspects for understanding the process of speciation. The relationship between genome regions and reproductive isolation traits in both structural and functional sense is gaining importance. GWAS, genome scans, QTL analysis, and study of chromosomal inversions are being utilized to identify genomic regions linked with traits affecting reproductive isolation (Hopkins et al., 2020; Wellenreuther and Bernatchez, 2018).

According to the recombination suppression model that explains the role of inversions in speciation, inversion suppresses recombination between loci for local adaptation and those for assortative mating, allowing divergence and speciation in the presence of gene flow (Huang and Rieseberg, 2020). Genetic structure rearrangements in the genome of sympatric sunflowers species (*Helanthus annuu* and *H. agrophyllus*) are linked with restricted gene flow and an inversion on chromosome 4 *A. thaliana* is strongly associated with fecundity under drought (Barb et al., 2014; Fransz et al., 2016). Inversions suppress recombination among blocks of linked genes, and as such, allow genes for selected traits and genes for reproductive isolation to be inherited together (Faria et al., 2019). Suppression of recombination in hybrids by inversion can help understand how the multiple genes that inversions contain help generate reproductive isolation. Suppression of gene expression or gene knockout of candidate loci with the CRISPR/Cas9 system can help to assess if a locus is linked with one or more traits impacting reproductive isolation (Hopkins et al., 2020).

8.9 ADAPTATION TO THE BIOTIC ENVIRONMENT

Interspecific or intraspecific interactions among species have positive or negative fitness effects. Plant enters into beneficial interactions with positive fitness effects that may help them in nutrition or defense against disease. On the other hand, interactions with herbivores and pathogens have an overall negative impact on fitness. However, some interactions are very complex, and as such, their effects are sometimes not so obvious. These biological interactions are considered a vital force for selection (Bamba et al., 2019; Finkel et al., 2017). Here is an overview of how genomic approaches have attempted to understand these interactions and unravel the signatures of adaptive phenotypes across the genomes and selection pressures acting on them.

8.10 PLANT–PATHOGEN INTERACTION

Plants in the natural environment are constantly faced with the challenge of herbivory and pathogen attack. In response, they have developed a multitude of defense mechanisms against herbivores and pathogens. These defense or resistance mechanisms in plants are essential adaptive responses that reduce the negative impact on fitness and help survival in the presence of pests and pathogens. *A. thaliana* is a model species used to understanding developmental, physiological and biochemical, genetic, and molecular aspects of plant life. It has also served as a model for genomic studies focused on understanding the genetic basis of adaptive responses and selection in various biotic and abiotic interactions. GWAS and genetic mapping have identified genes associated with disease resistance and unraveled complexities of trade-offs between adverse and positives fitness effects under different selection pressures. GWAS carried out on *A. thaliana* have identified disease resistance genes against *Hyaloperonospora parasitica* (a pathogenic fungus) and genes related to natural variations in disease resistance against *Pseudomonas syringae* (Nemri et al., 2010).

In plants, disease resistance proteins (R proteins) are intracellular receptors involved in pathogenic resistance. They activate effector-triggered immunity in plants after receiving and recognizing protein signal molecules of pathogenic origin (Petit and Fudal, 2017). R-genes, probably because of their higher energy costs, show a trade-off with growth and fitness (Brown and Rant, 2013). GWAS in *A. thaliana* revealed nonfunctional allele of the *ACQUIRED OSMOTOLERANCE* gene which is responsible for acquired osmotic stress tolerance. The functional allele confers resistance against bacteria in the absence of osmotic stress and reduces osmotic tolerance. It indicates a trade-off between biotic and abiotic stress, and the alleles are under selection (Ariga et al., 2017). *Fusarium* ear rot (FER) reduces grain yield and grain quality in maize. Wu et al. (2020) conducted a comprehensive study to understand natural variation in resistance against FER (caused by *Fusarium verticillioides)*. Mapping of QTL for resistance identified 10 QTLon 6 chromosomes, explaining 1.0–7.1% variation.

Further, epistatic mapping identified 4QTL pairs, explaining around 3% of the phenotypic variation.

8.11 HERBIVORY

Plants have evolved a multitude of adaptations at a morphological and biochemical level against herbivores. They have thorns, pubescence, and waxes to deter herbivores or produce toxic metabolites as a means of chemical defense against them. Glucosinolates are sulfur and nitrogen-containing secondary metabolites present in various plant groups, especially in *Brassicaceae*. They have a central carbon atom attached to a glucose group, a sulfate group, and a variable side chain. Variations in the side group are responsible for their varied biological roles. They are involved in plant defense against pests and pathogens. In plants, glucosinolates are hydrolyzed by myrosinase enzyme following tissue damage by the herbivore. The enzyme cleaves off the glucose group from the molecule, releasing biologically active products, such as isothiocyanates, nitriles, etc. which act as deterrents and toxins for insect pests (Fahey et al., 2001). In *A. thaliana,* GWAS and genome-wide selection scans of glucosinolate pathway have revealed that in the overall glucosinolate profile 36% among accession variation is explained by genetic differentiation only across three loci (Bamba et al., 2019; Brachi et al., 2015).

8.12 NUTRITIONAL SYMBIOSIS

Plants develop various types of interaction with bacteria and fungi for the sake of obtaining nutrition. For example, mycorrhizae are involved in the uptake of phosphorus and *Rhizobia* are interested in fixing nitrogen. The rhizobia inhabit root nodules of leguminous plants, and they fix atmospheric nitrogen into a usable form (Masson-Boivin et al., 2009). Genomic studies using genome-wide polymorphism data in *Medicago* have identified candidate SNPs associated with variation in nodulation. These are often found within or nearby genes responsible for nodulation and symbiosis (Stanton-Geddes et al., 2013). Disruption of gene function in candidate genes by various genomic techniques has helped identify genes that determine quantitative variation in symbiosis with rhizobia. Nucleotide diversity and gene expression data available for *M. truncatula* indicate the possibility of species-wide positives selection in genes for symbiosis (Bamba et al., 2019; Paape et al., 2013).

8.13 PLANT–MICROBIAL COMMUNITY INTERACTIONS

Plants interact with diverse groups of microorganisms, especially in the rhizosphere, creating a zone of extensive activity that facilitates the coexistence of the plant and microorganisms (Bulgarelli et al., 2013). These interactions with microbial communities benefit plants by promoting nutrient uptake, producing growth regulators, or protecting them against biotic and abiotic stress. Microbes colonizing the spermosphere and phyllosphere are also important as they impact seed germination, growth, and disease development in plants. Plant-associated microbial communities' composition varies with plant species, genotype, and stage of plant development and even with edaphic factors. Studies on microbial communities associated with *Arabidopsis*, maize, and *Populus* have revealed that host genotype has a significant influence on the microbes inhabiting the roots and phyllosphere (Bamba et al., 2019).

GWAS of leaf microbial communities in *A. thaliana* have shown that similar processes influence the leaves' fungal and bacterial communities. Gene ontology enrichment analysis revealed that "defense response," "cell wall integrity" and "trichome production" were over-represented across analyses (Horton et al., 2014). These studies indicate that plants' genetic architecture and microbial interaction are very complex and a better understanding of gene–gene and microbe–microbe interactions need to be assessed more thoroughly.

8.14 GENOMICS OF HYBRID ZONES AND SPECIATION

The term hybrid generally refers to offspring of a cross between parents of different species. If hybridization is considered reproduction between genetically different populations, it is involved in almost all types of speciation (Kliman, 2016; Counterman, 2016). Hybrid zones refer to geographic regions of contact or hybridization where related species mate to produce offsprings (Maloy and Hughes, 2013). "Hybrid zones occur when genetically distinct groups of individuals meet and mate, resulting in at least some offspring of mixed ancestry" (Harrison and Harrison, 1993).

Hybrids so produced in these zones of contact are sets of clines in populations' genetic composition from one taxon to another (Baldassarre et al., 2014; Saarman and Pogson, 2015).

Interspecific hybridization is considered an abrupt mechanism of speciation in higher plants, bringing together divergent genomes. It usually causes allopolyploidy, but rare homoploid speciation that occurs without genome duplication is also possible. Hybrids with odd ploidy numbers are either sterile or show reduced fertility. Genome stabilization in these hybrids may involve spontaneous chromosome doubling (allopolyploidy), or they may undergo introgression. Mostly allopolyploid hybrids are resistant to introgression (Petit et al., 1999), whereas homoploid hybrids are reproductively isolated from their parents by other mechanisms such as postzygotic isolation (Rieseberg, 1997). Species are identified as fertile groups between themselves and reproductively isolated from other such groups. Hybrids do not represent valid species unless they show reproductive isolation from parents and maintain their unique taxonomic identity (Ungerer et al., 1998). Grant (1978, 1981) identified two possible mechanisms for reproductive isolation of hybrids from their parents, leading to speciation. These are recombination speciation and transgression. The hybrids produced between species that differ by multiple chromosomal rearrangements are heterozygous for these rearrangements and partially sterile. Fifty percent viable gametes produced by them will have a parental type chromosome structure, and 50% will have a recombinant type. Events of inbreeding in these hybrids produce some offsprings with novel karyotypes that are fertile among themselves but to some extent, reproductively isolated from parental species (Rieseberg, 1997). Hybrid speciation can result from profound changes in genome and gene expression in hybrids that make them different from their parents in many aspects. Recent transcriptomic studies have shown differences in gene expression levels in interspecific hybrids as compared with their parents. Some of these are nonadditive while some are transgressive. In "transgressive segregation," new allele combinations in the hybrids produce "transgressive" phenotypes that help hybrids occupy ecological niches that neither parents could occupy (Hegarty et al., 2008; Rieseberg et al., 1999).

Study on natural and synthetic sunflower hybrids has shown that artificial combinations can survive in the extreme natural habitats where corresponding natural hybrid survive. It was reported that QTLs responsible for adaption to particular habitats were present in both natural hybrid and synthetic hybrids (Rieseberg et al., 2007). Analysis of genomic data comprising a large set of single nucleotide polymorphisms (SNPs) provides sufficient information to precisely estimate interspecific ancestry and the extent of introgression (Gompert et al., 2017). The formation of new species by interspecific hybridization in sympatry provides an opportunity to understand the speciation directly at the genome and transcriptome level (Hegarty et al., 2008).

8.15 CONCLUSION

In response to environmental changes, plants have evolved and developed more complex morphological and molecular aspects, facilitating them to adapt and survive better. The processes of adaptation and speciation are complex and interlinked. It is essential to understand how organisms detect and adapt physiologically to the changes in the environment. How diverse populations and species differ concerning their response to environmental changes. Genomics can identify the gain and loss of genes, changes in gene expression related to these novel traits, and unravels the genetic basis of the population's variable response to environmental changes. Speciation is an important process responsible for all the biodiversity that we have. Although there are many theories and hypotheses explaining speciation, none of these is foolproof. However, evolutionary biologists are invariably interested to understand the role of reproductive isolation, gene flow, hybridization, and local adaptation in the process of speciation. Genomic tools have opened new opportunities to identify the genomic regions associated with reproductive isolation and study the role of inversions in restricted gene flow. Although the genetic basis of local adaptation is very complex, genomic studies involving SNP's, GWAS, and genome scans have identified many candidate loci associated with adaptation to the local climate in plants. With genomic tools, it has been possible to ascertain profound changes in genome and gene expression in hybrids that make them different from their parents and allows them to inhabit novel habitats inaccessible to their parents. Novel tools like CRISPR/Cas9 system with gene editing ability have facilitated gene expression studies to find the specific role of candidate locus in reproductive isolation.

KEYWORDS

- **genomics**
- **adaptation**
- **speciation**
- **environment**
- **variation**
- **reproductive isolation**

REFERENCES

Abbott, R.; Albach, D.; Ansell, S.; Arntzen, J. W.; Baird, S. J.; Bierne, N.; Boughman, J.; Brelsford, A.; Buerkle, C. A.; Buggs, R. Hybridization and Speciation. *J. Evol. Biol.* **2012,** *26* (2), 229–246.

Abbott, R. J. Plant Speciation across Environmental Gradients and the Occurrence and Nature of Hybrid Zones. *J. Syst. Evol.* **2017,** *55* (4), 238–258.

Ariga, H.; Katori, T.; Tsuchimatsu, T.; Hirase, T.; Tajima, Y.; Parker, J. E.; Alcazar, R.; Koornneef, M.; Hoekenga, O.; Lipka, A. E. NLR Locus-Mediated Trade-off between Abiotic and Biotic Stress Adaptation in Arabidopsis. *Nat. Plants* **2017,** *3* (6), 1–8.

Bais, H. P.; Vepachedu, R.; Gilroy, S.; Callaway, R. M.; Vivanco, J. M. Allelopathy and Exotic Plant Invasion: From Molecules and Genes to Species Interactions. *Science* **2003,** *301* (5638), 1377–1380.

Baldassarre, D. T.; White, T. A.; Karubian, J.; Webster, M. S. Genomic and Morphological Analysis of a Semipermeable Avian Hybrid Zone Suggests Asymmetrical Introgression of a Sexual Signal. *Evolution* **2014,** *68* (9), 2644–2657.

Baldwin, I. T. Finally, Proof of Weapons of Mass Destruction. *Science's STKE* **2003,** *2003* (203), pe42–pe42.

Bamba, M.; Kawaguchi, Y. W.; Tsuchimatsu, T. Plant Adaptation and Speciation Studied by Population Genomic Approaches. *Dev. Growth Diff.* **2019,** *61* (1), 12–24.

Barb, J. G.; Bowers, J. E.; Renaut, S.; Rey, J. I.; Knapp, S. J.; Rieseberg, L. H.; Burke, J. M. Chromosomal Evolution and Patterns of Introgression in Helianthus. *Genetics* **2014,** *197* (3), 969–979.

Bouzid, M.; He, F.; Schmitz, G.; Häusler, R. E.; Weber, A. P. M.; Mettler-Altmann, T.; De Meaux, J. Arabidopsis Species Deploy Distinct Strategies to Cope with Drought Stress. *Ann. Bot.* **2019,** *124* (1), 27–40.

Brachi, B.; Meyer, C. G.; Villoutreix, R.; Platt, A.; Morton, T. C.; Roux, F.; Bergelson, J. Coselected Genes Determine Adaptive Variation in Herbivore Resistance Throughout the Native Range of *Arabidopsis thaliana*. *Proc. Natl. Acad. Sci.* **2015,** *112* (13), 4032–4037.

Brown, J. K. M.; Rant, J. C. Fitness Costs and Trade-Offs of Disease Resistance and Their Consequences for Breeding Arable Crops. *Plant Pathol.* **2013,** *62*, 83–95.

Bulgarelli, D.; Schlaeppi, K.; Spaepen, S.; Van Themaat, E. V. L.; Schulze-Lefert, P. Structure and Functions of the Bacterial Microbiota of Plants. *Annu. Rev. Plant Biol.* **2013,** *64*, 807–838.

Drabovich, A. P.; Martínez-Morillo, E.; Diamandis, E. P. Toward an Integrated Pipeline for Protein Biomarker Development. *Biochim. Biophys. Acta (BBA)–Proteins nd Proteom.* **2015,** *1854* (6), 677–686.

Fahey, J. W.; Zalcmann, A. T.; Talalay, P. The Chemical Diversity and Distribution of Glucosinolates and Isothiocyanates among Plants. *Phytochem.* **2001,** *56* (1), 5–51.

Fang, Z.; Pyhäjärvi, T.; Weber, A. L.; Dawe, R. K.; Glaubitz, J. C.; González, J. d. J. S.; Ross-Ibarra, C.; Doebley, J.; Morrell, P. L.; Ross-Ibarra, J. Megabase-Scale Inversion Polymorphism in the Wild Ancestor of Maize. *Genetics* **2012,** *191* (3), 883–894.

Faria, R.; Johannesson, K.; Butlin, R. K.; Westram, A. M. Evolving Inversions. *Trends Ecol. Evol.* **2019,** *34* (3), 239–248.

Finkel, O. M.; Castrillo, G.; Paredes, S. H.; González, I. S.; Dangl, J. L. Understanding and Exploiting Plant Beneficial Microbes. *Curr. Opin. Plant Biol.* **2017,** *38*, 155–163.

Fransz, P.; Linc, G.; Lee, C. R.; Aflitos, S. A.; Lasky, J. R.; Toomajian, C.; Ali, H.; Peters, J.; van Dam, P.; Ji, X. Molecular, Genetic and Evolutionary Analysis of a Paracentric Inversion in *Arabidopsis thaliana*. *Plant J.* **2016,** *88* (2), 159–178.

Ghanemi, A. Scientific Research in the Developing Countries: The Challenges We Need to Overcome. *MOJ Proteom. Bioinform* **2014,** *1* (3), 00017.

Gompert, Z.; Mandeville, E. G.; Buerkle, C. A. Analysis of Population Genomic Data from Hybrid Zones. *Annu. Rev. Ecol. Evol. Syst.* **2017,** *48*.

Grant, V. *Genetics of Flowering Plants*; Columbia University Press, 1978.

Grant, V. *Plant Speciation*; Columbia University Press, 1981.

Gregory, T. R. Understanding Natural Selection: Essential Concepts and Common Misconceptions. *Evol. Educ. Outreach* **2009,** *2* (2), 156–175.

Haak, D. C.; Kostyun, J. L.; Moyle, L. C. Merging Ecology and Genomics to Dissect Diversity in Wild Tomatoes and Their Relatives. *Ecol. Genom.* **2014,** 273–298.

Harrison, R. G.; Harrison, R. G. *Hybrid Zones and the Evolutionary Process*; Oxford University Press on Demand, 1993.

Hegarty, M. J.; Barker, G. L.; Brennan, A. C.; Edwards, K. J.; Abbott, R. J.; Hiscock, S. J. Changes to Gene Expression Associated with Hybrid Speciation in Plants: Further Insights from Transcriptomic Studies in Senecio. *Phil. Trans. R. Soc. B: Biol. Sci.* **2008,** *363* (1506), 3055–3069.

Hopkins, D. P.; Tyukmaeva, V. I.; Gompert, Z.; Feder, J.; Nosil, P. Functional Genomics Offers New Tests of Speciation Hypotheses. *Trends Ecol. Evol.* **2020,** *35* (11), 968–971.

Horton, M. W.; Bodenhausen, N.; Beilsmith, K.; Meng, D.; Muegge, B. D.; Subramanian, S.; Vetter, M. M.; Vilhjálmsson, B. J.; Nordborg, M.; Gordon, J. I. Genome-Wide Association Study of *Arabidopsis thaliana* Leaf Microbial Community. *Nat. Commun.* **2014,** *5* (1), 1–7.

Huang, K.; Rieseberg, L. H. Frequency, Origins, and Evolutionary Role of Chromosomal Inversions in Plants. *Front. Plant Sci.* **2020,** *11*, 296.

Kapun, M.; Fabian, D. K.; Goudet, J.; Flatt, T. Genomic Evidence for Adaptive Inversion Clines in Drosophila Melanogaster. *Mol. Biol. Evol.* **2016,** *33* (5), 1317–1336.

Kliman, R. M. *Encyclopedia of Evolutionary Biology*; Academic Press, 2016.

Kubota, S.; Iwasaki, T.; Hanada, K.; Nagano, A. J.; Fujiyama, A.; Toyoda, A.; Sugano, S.; Suzuki, Y.; Hikosaka, K.; Ito, M. A Genome Scan for Genes Underlying Microgeographic-Scale Local Adaptation in a Wild Arabidopsis Species. *PLoS Genet* **2015,** *11* (7), e1005361.
Lowry, D. B.; Willis, J. H. A Widespread Chromosomal Inversion Polymorphism Contributes to a Major Life-History Transition, Local Adaptation, and Reproductive Isolation. *PLoS Biol* **2010,** *8* (9), e1000500.
Ma, L.; Yang, L.; Zhao, J.; Wei, J.; Kong, X.; Wang, C.; Zhang, X.; Yang, Y.; Hu, X. Comparative Proteomic Analysis Reveals the Role of Hydrogen Sulfide in the Adaptation of the Alpine Plant Lamiophlomis Rotata to Altitude Gradient in the Northern Tibetan Plateau. *Planta* **2015,** *241* (4), 887–906.
Maan, M. E.; Seehausen, O. Ecology, Sexual Selection and Speciation. *Ecol. Lett.* **2011,** *14* (6), 591–602.
Maloy, S.; Hughes, K. *Brenner's Encyclopedia of Genetics*; Academic Press, 2013.
Manzoni, C.; Kia, D. A.; Vandrovcova, J.; Hardy, J.; Wood, N. W.; Lewis, P. A.; Ferrari, R. Genome, Transcriptome and Proteome: The Rise of Omics Data and Their Integration in Biomedical Sciences. *Briefings Bioinfor.* **2018,** *19* (2), 286–302.
Masson-Boivin, C.; Giraud, E.; Perret, X.; Batut, J. Establishing Nitrogen-Fixing Symbiosis with Legumes: How Many Rhizobium Recipes? *Trends Microbiol.* **2009,** *17* (10), 458–466.
Mayr, E. *Systematics and the Origin of Species*; Columbia UniversityPress: New York.
McKown, A. D.; Klápště, J.; Guy, R. D.; El-Kassaby, Y. A.; Mansfield, S. D. Ecological Genomics of Variation in Bud-Break Phenology and Mechanisms of Response to Climate Warming in Populus Trichocarpa. *New Phytol.* **2018,** *220* (1), 300–316.
Nemri, A.; Atwell, S.; Tarone, A. M.; Huang, Y. S.; Zhao, K.; Studholme, D. J.; Nordborg, M.; Jones, J. D. G. Genome-Wide Survey of Arabidopsis Natural Variation in Downy Mildew Resistance Using Combined Association and Linkage Mapping. *Proc. Natl. Acad. Sci.* **2010,** *107* (22), 10302–10307.
Nevo, E. "Evolution Canyon": A Microcosm Of Life's Evolution Focusing on Adaptation and Speciation. *Israel J. Ecol. Evol.* **2006,** *52* (3–4), 501–506.
Paape, T.; Bataillon, T.; Zhou, P.; Jy Kono, T.; Briskine, R.; Young, N. D.; Tiffin, P. Selection, Genome-Wide Fitness Effects and Evolutionary Rates in the Model Legume *Medicago truncatula*. *Mol. Ecol.* **2013,** *22* (13), 3525–3538.
Petit, C.; Bretagnolle, F.; Felber, F. Evolutionary Consequences of Diploid–Polyploid Hybrid Zones in Wild Species. *Trends Ecol. Evol.* **1999,** *14* (8), 306–311.
Petit, Y.; Fudal, I. Complex Interactions between Fungal Avirulence Genes and Their Corresponding Plant Resistance Genes and Consequences for Disease Resistance Management. *Front. Plant Sci.* **2017,** *8*, np.
Pinho, C.; Hey, J. Divergence with Gene Flow: Models and Data. *Annu. Rev. Ecol. Evol. Syst.* **2010,** *41*, 215–230.
Rieseberg, L. H. Hybrid Origins of Plant Species. *Annu. Review Ecol. Syst.* **1997,** *28* (1), 359–389.
Rieseberg, L. H.; Kim, S.-C.; Randell, R. A.; Whitney, K. D.; Gross, B. L.; Lexer, C.; Clay, K. Hybridization and the Colonization of Novel Habitats by Annual Sunflowers. *Genetica* **2007,** *129* (2), 149–165.

Rieseberg, L. H.; Whitton, J.; Gardner, K. Hybrid Zones and the Genetic Architecture of a Barrier to Gene Flow between Two Sunflower Species. *Genet.* **1999,** *152* (2), 713–727.

Saarman, N. P.; Pogson, G. H. Introgression between Invasive and Native Blue Mussels (Genus Mytilus) in the Central California Hybrid Zone. *Mol. Ecol.* **2015,** *24* (18), 4723–4738.

Sato, Y.; Shimizu-Inatsugi, R.; Yamazaki, M.; Shimizu, K. K.; Nagano, A. J. Plant Trichomes and a Single Gene GLABRA1 Contribute to Insect Community Composition on Field-Grown *Arabidopsis thaliana. BMC Plant Biol.* **2019,** *19* (1), 1–12.

Savolainen, O.; Lascoux, M.; Merilä, J. Ecological Genomics of Local Adaptation. *Nat. Rev. Genet.* **2013,** *14* (11), 807–820.

Schluter, D.; Conte, G. L. Genetics and Ecological Speciation. *Proc. Natl. Acad. Sci.* **2009,** *106* (Suppl 1), 9955–9962.

Stanton-Geddes, J.; Paape, T.; Epstein, B.; Briskine, R.; Yoder, J.; Mudge, J.; Bharti, A. K.; Farmer, A. D.; Zhou, P.; Denny, R. Candidate Genes and Genetic Architecture of Symbiotic and Agronomic Traits Revealed by Whole-Genome, Sequence-Based Association Genetics in Medicago Truncatula. *PloS One* **2013,** *8* (5), e65688.

Tarazona, E.; Hahn, C.; Franch-Gras, L.; García-Roger, E. M.; Carmona, M. J.; Gómez, A. Ecological Genomics of Adaptation to Unpredictability in Experimental Rotifer Populations. *Sci. Rep.* **2019,** *9* (1), 1–11.

Ungerer, M. C.; Baird, S. J. E.; Pan, J.; Rieseberg, L. H. Rapid Hybrid Speciation in Wild Sunflowers. *Proc. Natl. Acad. Sci.* **1998,** *95* (20), 11757–11762.

Ungerer, M. C.; Johnson, L. C.; Herman, M. A. Ecological Genomics: Understanding Gene and Genome Function in the Natural Environment. *Heredity* **2008,** *100* (2), 178–183.

Voelckel, C.; Gruenheit, N.; Lockhart, P. Evolutionary Transcriptomics and Proteomics: Insight into Plant Adaptation. *Trends Plant Sci.* **2017,** *22* (6), 462–471.

Wellenreuther, M.; Bernatchez, L. Eco-Evolutionary Genomics of Chromosomal Inversions. *Trends Ecol. Evol.* **2018,** *33* (6), 427–440.

Wu, Y.; Zhou, Z.; Dong, C.; Chen, J.; Ding, J.; Zhang, X.; Mu, C.; Chen, Y.; Li, X.; Li, H. Linkage Mapping and Genome-Wide Association Study Reveals Conservative QTL and Candidate Genes for Fusarium Rot Resistance in Maize. *BMC Genom.* **2020,** *21*, 1–11.

Xu, Q.; Zhu, C.; Fan, Y.; Song, Z.; Xing, S.; Liu, W.; Yan, J.; Sang, T. Population Transcriptomics Uncovers the Regulation of Gene Expression Variation in Adaptation to Changing Environment. *Sci. Rep.* **2016,** *6* (1), 1–10.

Yoder, J. B.; Stanton-Geddes, J.; Zhou, P.; Briskine, R.; Young, N. D.; Tiffin, P. Genomic Signature of Adaptation to Climate in Medicago Truncatula. *Genetics* **2014,** *196* (4), 1263–1275.

Zapalska-Sozoniuk, M.; Chrobak, L.; Kowalczyk, K.; Kankofer, M. Is it useful to Use Several "Omics" for Obtaining Valuable Results? *Mol. Biol. Rep.* **2019,** *46* (3), 3597–3606.

Zhang, M.; Suren, H.; Holliday, J. A. Phenotypic and Genomic Local Adaptation Across Latitude and Altitude in Populus Trichocarpa. *Genome Biol. Evol.* **2019,** *11* (8), 2256–2272.

CHAPTER 9

Role of Ecogenomics in Conservation and Management

SHAHID UL ISLAM[1], ZAHID AHMAD MANGRAL[1], LUBNA TARIQ[1], BASHARAT AHMAD BHAT[2], HAIDER MEHRAJ[3], SHEIKH TAJAMUL ISLAM[2], and TANVIR UL HASSAN DAR[1,*]

[1]*Department of Biotechnology, School of Biosciences and Biotechnology, BGSB University, Rajouri, Jammu and Kashmir, India*

[2]*Department of Bioresources, University of Kashmir, Srinagar, Jammu and Kashmir, India*

[3]*Department of Electronics and Communication Engineering, School of Engineering and Technology, BGSB University, Rajouri, Jammu and Kashmir, India*

[*]*Corresponding author. E-mail: tanvirulhasan@gmail.com*

ABSTRACT

Ecogenomics is the scientific approach to understand the relationship between structural and functional aspects of genomes with biotic/abiotic environmental factors. The classification of ecology depends upon the overall complexity (behavioral and population ecology) of an organism (plant and animal ecology) and system under investigation (soil and forest ecology). The molecular techniques adopted by these various ecology branches result in a new field known as ecological genomics or ecogenomics that focuses on an organism's overall development during the evolutionary period. It is an interdisciplinary research field covering ecological science, microbiology,

Plant Ecogenomics. Peerzada Arshid Shabir, PhD, Peerzada Yasir Yousuf, PhD & Khalid Rehman Hakeem, PhD (Eds.)

environmental and molecular biology, toxicology, physiology, chemistry, etc. The study related to the ecology of plants has a direct relationship with the adaptation mechanism because plants have no alternative to cope with the environment in which they grow. The world is facing biodiversity loss at an alarming rate, with a loss of 90% crop varieties in a century from the field. It is essential to select the genes in biological pathways responsible for an organism's stability in the ecological system. The challenge is to understand the basic phenomenon behind adaptation, migration, inbreeding mechanism of endangered or critically endangered species. In ecogenomics, we know that the genes are responsible for the effective management strategies from a conservation point of view. Therefore, in this chapter, we discuss various aspects of community shaping and visualization. The community's function and structure need to be studied due to the availability of plenty of molecular data. Further, the role of ecogenomics and multiomic approaches in conservation and management will also be emphasized.

9.1 INTRODUCTION

Ecogenomics is the relationship between organisms and their dynamic environment. We can also define it as the relationship between an organism's ecology and molecular level interactions (Thomas and Klaper, 2004; Van Straalen and Roelofs, 2006). It covers all aspects of ecology and genomics by studying biochemistry, molecular and cell biology, and population biology, leading to an organism's ecosystem and biosphere (Handelsman et al., 1998; Maphosa et al., 2010). The classification of ecology depends upon the overall complexity (behavioral and population ecology) of an organism (plant and animal ecology) and system under investigation (soil and forest ecology). The molecular techniques adopted by these various ecology branches result in a new field known as ecological genomics or ecogenomics that focuses on an organism's overall development during the evolutionary period. We can easily understand the effect of the change made to an organism by inserting a gene of interest (Baldwin et al., 2006; Schranz et al., 2009). One can speak about this field only when ecology and molecular biology are studied together (Chariton et al., 2010; Maphosa et al., 2010). It also involves studying an organism concerning its surroundings to understand its role, existence, and impact on other associated species. The genes in nature that are responsible for the decomposition of a particular substance unfavorable to the environment can be inserted into an organism. So that it could protect the environment being affected by the hazardous nature of that substance (Ozean, 2012).

Not all the researchers follow this approach in this field. Others characterize this as the bulk of genetic material present in the given environment and are wrongly associated with metagenomics (Kakirde et al., 2010). Almost everyone claims that ecogenomics plays a role beyond genetics at a lower or higher level. Its main aim is to generate perception regarding the relationship between living entities and their environment by applying various molecular tools. (Fig. 9.1).

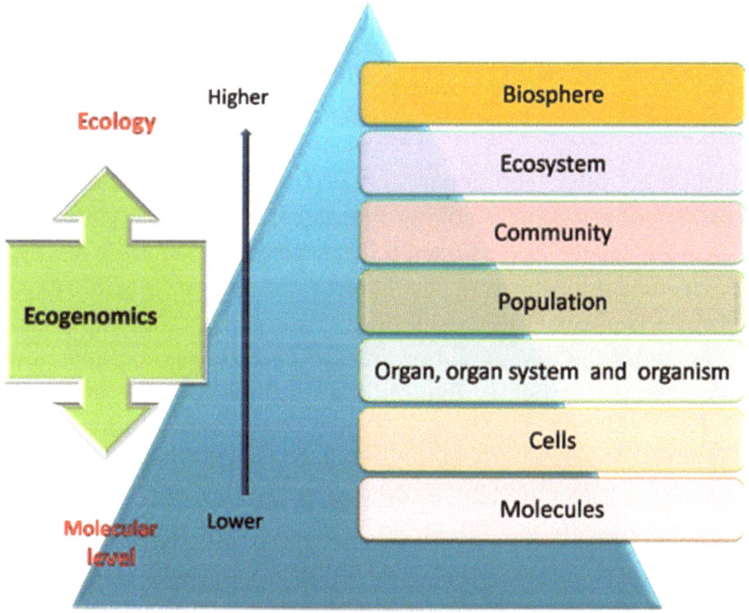

FIGURE 9.1 Systematic representation of ecogenomic approach.

According to Fox Keller, the 20th century is considered the century of genes. After discovering laws of inheritance by George Mandel published in 1865, these laws were rediscovered by Devries, Correns, and Von Tschermak, which became the landmark for the century of genes. These laws revolutionized this era in biology and gave rise to genetics in 1902, the genome in 1909, genomics in 1987, and ended with the complete human genome sequencing in 2000 (McKusick and Ruddle, 1987). This revolution impacted genetics in the present scenario. Even one cannot think about research and other processes without genes and genomics. The mixture of ecology and genomics has led to the latest and vast ecological and evolutionary functional genomics in an ecological context, that is, ecological genomics fulfills ecology's primary purpose. In contrast, genomics is the modern method to address the fundamental issues related to ecology. Genomics and ecology are in a dearth of the need of tools that are the reason for the new scientific questions raised. New technologies with current problems can be addressed in a way that has not been previously considered.

9.2 AIMS OF ECOGENOMICS

The ecological study has a long history of studying adaptation because the individual plants have no alternative but to cope with the environment in which they grow. Evolutionary research studied the adaptation of the interaction between evolutionary forces (i.e., more specifically, migration and drift) and selection, resulting in transformation. The use of genomic tools/techniques now helps open a black box between the phenotype and the environment to illustrate the adaptation mechanisms while examining phenotypic changes. Variation in fitness and phenotypes can provide the base for understanding evolution. It also investigates the genetic targets for the selection process. Evolutionary influences exist at the genetic level, but the focus of selection is on phenotypes. The introduction of genomic techniques/tools may get used to classify emerging genes (Lee and Mitchell, 2006). It is also pertinent to mention here that the research process will provide the opportunity to understand evolutionary complex phenomena, that is, genetic drawbacks and the origin of phenotypic plasticity (Jackson et al., 2002).

9.3 ECOGENOMIC METHODS USED IN CONSERVATION AND MANAGEMENT

The number of tools that genomic research provides for conservation and management is huge and still increasing. There are three broad approaches of ecogenomics to explore the ecological functioning of organisms:

(i) Variation in Sequences
(ii) Expression Analysis
(iii) Gene Function Analysis

9.3.1 VARIATION IN SEQUENCES

This type of variation is determined based on the sequences in the DNA molecule and most of the integrated ecology techniques are used to detect these variations. Most of the methods have been developed to facilitate interpretation in the regions of certain genes or gene-promoter. Sequencing information is used in two different ways in conservation and management. Firstly, simple sequence repeat and single nucleotide polymorphism, also known as SSR and SNPs markers, can be used for selection studies. As compared with SSR and AFLPs, SNPs contain a large part of the genome and form an excellent marker to study the variation particularity in the regions which are coding in nature (Lee et al., 2004). These markers are used to detect genetic variation within various life forms and depict information related to inbreeding, gene flow, genetic diversity, etc. Second, they are beneficial in studying the footprints leading to selection.

9.3.2 EXPRESSION ANALYSIS

The change of gene expression in trans or cis-regulatory mechanisms resulting from genetic drift and inbreeding are mostly unknown. Evolutionary changes based on the phenotypic diversity among and within the species are governed by the gene expression, which is considered the ground for the phenotypic evolution (Meiklejohn. et al., 2014). There has always been an argument regarding gene expression and the morphological variation in the plant species (Khaitovich, et al., 2006; Nielsen, et al., 2009). As it is evident that abiotic stress showed a positive and negative effect on both plants and animals. The increasing temperature and food scarcity is one example, which results in the change of genomic pathways, both in animals and plants (Sorensen et al., 2007). In *Drosophila melanogaster*, significant gene expression profiles observed in inbred lines compared with outbred lines (Kristensen et al., 2005). We can assess how genes are expressed, particularly concerning the gene transcript concentration estimation in a specific tissue at a particular time. The dissimilarity between the samples at transcription levels is identified because they play an essential role in adaptation behavior. When the information related to the sequence is available in bulk, either by cDNA or microarrays, gene expression variation (transcriptome analysis). A microarray collects microscopic DNA samples attached to a solid surface (glass or quartz surfaces) developed by high-speed robotics. Microarray technology is of two major types that vary in their DNA properties. Likewise, a PCR-based technique DNA-AFLP (amplified fragment length polymorphisms) and qRT-PCR (quantitative real-time PCR) may be utilized for analysis at transcriptome level even though if no sequence data are available. These methods of gene expression are being used to classify evolutionarily crucial genetic variation and are briefly described below:

9.3.2.1 cDNA MICROARRAY (DNA CHIPS)

Microarray is one of the best techniques to examine the change in ERs (expression ratios) involved in different organisms. This technique can elucidate the beginning stages of a phenotypic response in an organism, leading to ecologically essential interactions. An example is the transcriptomic level changes in response to the arsenate observed by the cDNA-microarray analysis. The genes that are differentially expressed were identified for the biomarker identification after identifying three biomarker genes for dose-dependent upregulation (Brinke et al., 2018). This technique thereby revolutionized most biology and is also useful for studying plant and herbivore interactions (Held et al., 2004). Researchers also documented that the plants with various stresses, such as drought, ozone stressed, or wounding result in plant growth inhibition. However, the factors that caused damage to the plants are not visually detected or identified. For the plant's diagnosis and conservation, a reliable method is developed called cDNA microarray to measure the gene expression changes between stress by drought, ozone, or wounding. This analysis was utilized to measure the expression in the gene that occurs in *Arabidopsis thaliana*, caused by ozone stress (Matsuyama et al., 2002). Sanchita and Sharma in 2018 presented a detailed account of the gene expression analysis, which occurs in medicinal plants due to various environmental stresses. They concluded that these stresses are very complicated, reflecting their effect at morphological, biochemical, and physiological levels. cDNA microarray (commonly known as DNA chip or biochip) is a new and most powerful technique used for studying gene expression in different organisms. DNA samples may be genomic DNA, cDNA, or PCR product coated on a solid surface and hybridized to several selected points in this technique. These selected points are formulated from mRNA, and the cDNA thus formed are tagged with a fluorescent dye (Schena et al., 1995). The fluorescent dyes used are Cyanine 5-dNTP and green Cyanine 3-dNTP. The fluorescently labeled RNAs are hybridized onto a single glass slide for imaging. Differences in the mRNA between these samples can easily be seen as different colors of the spot. These color variations clearly show the magnitude of expression of that particular gene among the two models. The best example in recent time is the RNA scope RNA-ISH method that can solve the problem by novel single amplification. By this method, we can recognize various targeted RNAs in a single molecule of an individual, which is beneficial for the study of gene expression at spatioterminal patterning. In barley, the distribution pattern of HvGAPDH and low expression of Rpg1 (disease resistance gene) enables the researcher to discuss the precaution in the future (Solanki, 2020).

9.3.2.2 OLIGONUCLEOTIDE MICROARRAY

Oligonucleotide chips are another type of microarrays; the basic idea in these chips is like DNA microarray. These chips contain the oligos and not the targets. These are glass slide arrays generated by the photolithographic technique that enabling direct synthesis of the oligonucleotide DNA sample quartz wafer. The size of the probes used in this array is the range of 25–60 base pairs, and "ordinary chips" for many plants like *Arabidopsis thaliana*, corn, barley, tomato, and soybean are commercially available examples. These chips are manufactured in such a way as to reject the target that is nonidentical. We need specificity to express the sequences accurately, which is expressed in two similar genes. These chips are usually more sensitive than cDNA chips and provide the best signals for gene hybridization. It is important to note that these chips' production needs information about the DNA sequences that are to be identified. Therefore, ESTs (expressed sequence tags) and sequencing projects of plant genomes are crucial for developing oligonucleotide chips. These chips must come from a quality manufacturer, whereas the in-house company develops cDNA microarrays. Kyselková and co-workers (2009) concentrate on the study that this technique is used in microbial ecology to identify microorganisms in complex environmental samples. Analysis of microarrays has been reported as a critical approach to explaining the functional and taxonomic microbial communities' composition in environments, such as compost, soil, air, human, etc.

9.3.2.3 cDNA-AFLP

cDNA-AFLP (amplified fragment length polymorphisms) another method that does not need DNA sequence information widely used in various plants for the genome-wide expression analysis (Breyne and Zabeau, 2001; Cnuddeet al., 2003; Bae et al., 2006). Studies also revealed that the fungal pathogen *Puccinia striiformis* sp. *tritici* caused stripe rust, one of the world's acute diseases. One more example is a pathogen on the wheat host. cDNA-AFLP is a versatile tool used for analyzing gene expression involved in the interaction of wheat and fungal pathogen (Wang, et al., 2009). It is a PCR-based approach for determining gene expression profiles using band intensity (Breyne et al., 2003). Apart from this, it is peculiar and sensitive to detect weak expressed genes and minute variations in the activities that occur in the transcription.

cDNA-AFLP (amplified fragment length polymorphisms) method begins with the reverse transcription of mRNA. This step is followed by restriction digestion of cDNA with two separate restriction enzymes and the ligation of specific adapters to both ends of the fragments to provide known sequences for the PCR amplification. The PCR amplifies only those fragments which are digested by using restriction enzymes, and the two separate adapters are involved at both ends of the digested fragment. The mixture of cDNA is then subjected to selective PCR amplification and divided into smaller pieces with adapters-based primers in the following steps. With more selective nucleotides, selective primers are radioactively labeled, enabling the fragments to be visualized on a high-resolution gel (agarose gel/polyacrylamide gel with autoradiography). The fragments having interesting intensity patterns are isolated from the gel and sequenced. Thus, both cDNA-AFLP and microarray methods are used to detect/predict genes based on expression patterns. Dong in 2016 analyzed the gene expression in cabbage and *Hyaloperonospora,* which revealed that the effect of the resistant genes is high in the resistance level compared with the susceptible one, which is the main reason for the resistance against the pathogens. It will lead to crop improvement (Xiao, 2016). One more technique known as SSH PCR (subtractive hybridization polymerase chain reaction) is cDNA subtraction based on genes enrichment precisely the gene expression in one and elimination in both sample tissues on the comparison. It may be useful for the isolation of enabled genes even though this approach is not appropriate for identifying the gene modulated in expression. In particular circumstance (Siebert et al., 1995), these two techniques, SSH and cDNA-AFLP, widely help in identifying several expressed gene which is responsible for several variations.

9.3.2.4 qRT-PCR

The qRT-PCR (quantitative real-time reverse transcription polymerase chain reaction) is a technique that involves the general principle of PCR technique where one can amplify and simultaneously quantify a target DNA molecule. The nucleic acid concentration can be quantified initially using a fluorescent detection system measured with elevation in the exponential PCR phase (Heid et al., 1996). Generally, microarrays and qRT-PCR findings are closely correlated (even though there are sometimes visible and notable exceptions) (Dallas et al., 2005). This technique is technically less challenging but can examine a few genes at each interval on the screen for their amplification which is not possible in the microarray. Antonella and Luca (2013), suggested the qRT-PCR technique as one of the molecular methods for accurately identifying and characterizing marine toxic microalgal species. RT-PCR technology regulates and monitors the critical group that produces biotoxins and harms the marine environment, human health, and related economic activities.

9.3.3 GENE FUNCTION ANALYSIS

Gene functions are very important to know how various processes are being organized, formed, and operated by the plant. Genes are different in their position, and it is crucial to define them in a suitable manner (Hansen, 2014). A modification that occurs in the gene is the reason for various variants known as an allele. In plants, these modifications in terms of gene expression result in a change in the phenotype and put more focus on its role, and this can be accomplished through a variety of different approaches.

9.3.3.1 GENE FUNCTION BY MUTAGENESIS

Gene can be mutated not to generate a working protein. Its expression can be modified (activated or silenced). Mutations are incorporated in many ways either biologically with the help of transposons or chemically using radiation or other approaches, leading to various mutations (Alonso and Ecker, 2006). Gene function can be established through mutagenesis in different organisms, as discussed by several researchers in *Arabidopsis thaliana* (Ostergaard and Yanofsky, 2004). For example, EMS (ethyl methane sulpfonate) is a mutagen responsible for mutation at a single base-pair. TILLING (targeted induced local lesions in genomes), a method, is used to produce modifications for high throughput sequencing and applied to various plant species, for example, *Arabidopsis thaliana* and *Brassica oleracea* (Colbert et al., 2001; Till et al., 2003). Multiple forms of polymorphism are identified in natural populations by Ecotilling and are efficient in determining genetic variation in a wide variety of genotypes of any plant species (Comai et al., 2004).

9.3.3.2 VIRUS-INDUCED GENE SILENCING

Virus-induced gene silencing also referred to as VIGS, is a powerful technique used for plant functional genomics. Another method called gene silencing can also be used for gene function analysis without producing a mutant. It is one way to build knockout gene phenotypes (Robertson, 2004) efficiently. It is an RNA-based gene silencing that implies generating large amounts of dsRNA, which initiates mRNA silencing complementary to dsRNA sequences. In this method, dsRNAs are developed by using some RNA viruses to activate the silencing mechanism. The modification in the virus is carried out so that the sequences produced for the gene of interest in plants result in mRNA degradation from that gene. This mechanism, RNAi (RNA interference), most probably known as co-suppression or posttranscriptional gene silencing (PTGS), are used as a protective mechanism against endogenous transposons and viruses (Fire et al., 1998; Waterhouse and Wang, 2001; Robertson, 2004). For versatile reverse genetic tools, plant virus vectors are used for practical, functional analysis. These plant viruses express the genes heterologously and can suppress the gene expression by VIGS. The best example for understanding this is that this method is used in these plants which are not susceptible to *Agrobacterium*. This kind of study was conducted in the *Saline latifolia* plant (Hu et al., 2015). In another study, Fujita et al. 2019, demonstrated that when *Saline latifolia* infected 100% with ALSV (Apple latent spherical virus), the infected plant showed no symptoms while flower meristems were full of virus vectors. The knockdown of the phytoene desaturase gene, SIUFO and SISUP is confirmed by VIGS essential for observing phenotypes of the flower. Gene silencing is also helpful in creating artificial resistance in plants as they show resistance against diseases, such as virus infections, plant nematodes, crown gall disease, disease-resistant rootstock, etc. (Ebelechukwu et al., 2014).

9.3.3.3 *AGROBACTERIUM TUMEFACIENS METHOD*

This method is also used for studying the function of the gene by incorporating foreign DNA into the plant genome. *Agrobacterium tumefaciens* is the most used method to include a vector possessing DNA that can activate or silence a particular gene under observation. It also enables DNA transfer to the plants for RNAi and VIGS. Unfortunately, some species are resistant to *A. tumefaciens*, *Rhizobium* spp., etc., which may seriously reduce transgenic plants' generation. The study of genes' effect in the GMOs (genetically modified organisms) can be best checked in natural conditions. However, a big issue with these modified organisms is how they inhabit the ecological niches. Since it cannot be excluded, the wild organisms can compete and generate hybrids, but GMOs can only be tested with controlled conditions. This method is widely used in the ecogenomic approach for various interactions between the bacteria and the plants which is useful for enhancing the protein expression in the plants, plant transformation concerning the disease resistance, or any other effect beneficial for the plants (Hwang, 2017).

9.3.3.4 *QUANTITATIVE TRAIT LOCUS*

Quantitative trait locus (QTL) is referred to as a particular piece of DNA region having a specific phenotypic trait. QTLs are usually present or found on different chromosomes. It is characterized as a genetic locus, where alleles with functional differences result in a significant change/effect on the quantitative trait. QTL mapping studies and the information on a specific feature in different conditions lead to identifying genes responsible for polymorphism. If genome sequencing has been done in individual cases, it is feasible to find the QTL on the sequence and then specifically search for desired genes (Price, 2006). The plant disease's resistance is studied by QTL mapping, which is a highly successful method. The loci with specific resistance functions are explained by using QTL mapping. In addition, other studies can be carried out, such as resistance genes interactions, a race specificity of partial resistance genes can be assessed, interactions between plant growth, environment. Examples include late blight of potato, gray leaf spot in maize, rice blast fungus, a nematode of soybean cyst, and bacterial wilt of tomato (Young, 1996).

9.4 IMPORTANCE OF ECOGENOMICS IN SHAPING COMMUNITIES

Ecosystem structure and function are the two main components that play an essential role in shaping the community. The structure of an ecosystem comprises species diversity, biomass and dominant structure, and feeder groups. Simultaneously, the functional part consists of primary development, respiration, deterioration of organic matter, nitrification, etc. these processes cannot be visualized but can be tracked over time. During the 1990s, ecologists were curious to know about these two main components. There is an uneven relationship between these two components, that is, the ecosystem's function depends on all the available structures (Domsch, 1984). Generally, a minimum number of species is required for smooth functioning, while a vast number is needed for an ecosystem's stability (Loreau et al., 2001). Two points are essential which show how the problem can be benefited from this issue. Firstly, how the diversity of an ecosystem affects ecologically important processes depends on how restricted they are. This problem is of obvious importance in aquatic environments, where primary production may be limited either by an organism's ability in substrate processing. The substrate is quickly supplied in the system, where the capacity is limited until each functional unit gets saturated. The rate of output gets changed as per the speed of the supply of the substrate. An example is the nitrogen fixation in cyanobacteria, where the nitrogen gas does not affect the fixation, but the biomass of the fixing species execute. Secondly, there is a thought that biodiversity is not as important as its relationship with the organisms' characteristics. To gauge the impact of dwindling species abundance on ecosystem systems, we need to look at the biodiversity of the community's ecological character, not just the species number. Biodiversity and the ecosystem's working synthetic structure must be studied using the genomic method because a great deal of mechanistic understanding has increased exponentially through transcription and large-scale sequencing.

It is essential to understand the developmental process possible because of ecology and genomics, which entirely depends on the species nature and environment. For example, the genes involved in the primary functioning of species to tackle the environment. The diversity in genes can be taken into account without organisms in which they are. They can be used as indicators of the specific function involved in an ecosystem. Few genomic studies that are being performed to respond to the positive link between the overall process and biodiversity of the species underpin the rivet hypothesis related to biodiversity and the ecological. The dominant member of the population does not always perform the ecosystem's primary function, as suggested by population genetics studies. The transcriptome analysis can assess better environmental functioning, where the extraction of mRNAs is done in the eukaryotic model. It is also restricted to the profiling of pure cultures rather than on-the-ground societies. However, there are some exciting recent technological advances in which whole-community RNA amplification is used to extract adequate microbial mRNA quantities from environmental samples (Gao et al., 2007). Metagenomics has created recent attention for bio-prospective research. The analysis of novel functions employing metagenomic libraries in which unknown genes are expressed in the environment's heterogeneous hosts. This strategy has so far mainly centered on biotechnological significance, but we see no reason why it should happen in the same way. Recent findings have also proved that spatial dimensions and geographical barriers significantly affect forming cultures previously thought. Sources of biodiversity should be assessed to evaluate the relationships between structure and function for community shaping (Curtis and Sloan, 2004).

9.5 GENETIC LEVEL VARIATION FOR CONSERVATION

Due to genetic drift and inbreeding, isolated and small populaces of species are vulnerable to genetic diversity erosion. This loss of genetic variability will reduce adaptation's revolutionary potential in changing environments and causes inbreeding depression. Along with other population-dependent stochastic processes, this could increase the probability of population extinction. This collection of methods and hypotheses makes the foundation of conservation genetics. It is the study of genetics to understand and decrease the risk of population and species extinctions. Conservation genetic experiments with DNA markers have proved to help detect hereditary loss, reconstructing population history, and quantifying and identifying different processes at the landscape level. Moreover, the use of these DNA markers, which are now commonly used, also has an offensive and unsought side. It is always tempting to consider marker data as entirely predictive of fitness and extinction possibilities. There are three forms of genetic variation (neutral, negative, and adaptive) with particular concepts and consequences regarding conservation.

9.5.1 NEUTRAL VARIATIONS

Neutral genetic variation in a population is mainly controlled by genetic drift, mutation, migration, and recombination during interactions (Fay and Wu, 2003), while as the migration results in the loss of genetic diversity, and genetic drift, mutation, and recombination result in the new variants to the existing ones. Genetic conservation is essential depending on small-scale neutral marker surveys, particularly in historical gene flow inferences that help assess populations, specificity, and conservation status. Small amounts of neutral variation commonly create questions about the loss of adaptive abilities, inbreeding depression, and genetic loading (Hedrick, 2001; Kirkpatrick and Jarne, 2000) (Table 9.1). However, few neutral markers variation could conceptually expect dangerous and adaptive variation levels. Available literature reveals that the neutral variation is not closely connected with detrimental and adaptive variations and neglects population heterogeneity in quantitative characters (Hedrick, 2001; Reed and Frankham, 2003; Merila and Crnokrak, 2001). The use of neutral marker surveys in a limited manner can lead to misconceptions in terms of adaptive and detrimental variation in endangered species.

TABLE 9.1 Three Forms of Genetic Variations with Conceptual Consequences for Conservation: ++ (High),−(Low), and −/+ (Low to high).

Population structure		**Variation**		
		Observed	**Predicted**	
		Neutral variation	**Detrimental variation**	**Adaptive variation**
Equilibrium	Large population	++	++	++
	Small population	–	–	−/+
Bottleneck	Shortly after	–	–	−/+
	Sometime after		++	−/+
	Separately bottlenecked populations (mixture)	++	–	−/+
Metapopulation	Low gene flow and Extinction recolonization dynamics	–	–	−/+
	No extinction and substantial gene flow	++	++	++

9.5.2 DETRIMENTAL VARIATION

In this variation, gene sequences of proteins are abnormally changed, influencing an individual (Sunyaev et al., 2001; Reumers et al., 2005). These variations vary from population to population, that is, variations are more effective in large than in small populations. The minor effect due to detrimental variation is negligible in the populations and has a low frequency, resulting in low genetic load (Crow, 1999; Kondrashov, 1995; Keightley and Eyre-Walker, 2000; Ohta, 1992). Minor effects in a population are ranged as $^1/_2N < s < ^1/_2N$ (where N =genetically sufficient population size and s= selection coefficient), where the genetic drift increases the frequency if the effect is neutral. The increase in the small population results in a decrease of the mean fitness that directly affects the mutation exhibited by an individual (Gaggiotti, 2003; Spielman et al., 2004). This mutation can be governed by the increased homozygosity, which is directly proportional to the inbreeding depression (Hedrick, 2001; Sousa et al., 2000). The population structure suggests that detrimental mutations are likely to be swept out when the population size decreases (Table 9.1) (Hedrick, 2001; Kirkpatrick and Jarne, 2000). Ecogenomics helps in population control so that these variations should be minimal or negligible to reduce the genetic load and decrease inbreeding depression (Woodworth et al., 2002).

9.5.3 ADAPTIVE VARIATIONS

Genetic variation helps to respond to environmental challenges within a population, an example is the changing climate (Frankham 2003; Lande and Shannon, 1996). Reduction in the population size is the reason behind the simultaneous decreases in adaptive variation. Minor variations at the genetic level are retained in small populations (Hedrick, 2001). Simultaneously, the chances of recombination are minimized with some new variations that occur per unit time. Beneficial new mutations rarely occur even in large populations (Fay and Wu, 2003). The variations can be preserved by selection reduced in adaptive variations, while choice does not affect low detrimental variations (Hedrick, 2001). Conservation strategy, genetic adaptation in a population, and adaptive divergences (ecology and life-history traits) are essential in maintaining adaptive genetic variation. As a result, the ability to adapt weakly adaptive alleles in endangered species is reduced while reflect adaptive variations may hypothetically be targeted for conservation (Lande and Shannon, 1996; Crandall et al., 2000; Reed and Frankham, 2003: Rader et al., 2005) (Table 9.1).

9.6 ROLE OF ECOGENOMICS IN CONSERVATION MANAGEMENT

The population's fitness is improved by introducing new variations at the sequence level, which is done by incorporating individuals from closely associated populations to an affected population. The result is evident in the small and isolated populations suffering from inbreeding depression (Frankham, 2015). The increased genetic load reduces the efficiency in the selection process, which increases the genetic drift of alleles to the high frequency. This process is covered under one heading, that is, genetic rescue, where the genetic load decreases with introducing new variants with desirable traits (Hedrick and Fredrickson, 2010). This genetic rescue program is highly effective from the conservation point of view because it helps overcome the inbreeding depression and the problems faced in predicting the proposed outcomes. Population size and demographic background are the primary tools for understanding the degree of inbreeding and deciding genetic rescue in that particular population. The degree of inbreeding is measured genome-wide commonly by using a genetic marker which is more reliable, and pedigrees predict inbreeding (Kardos et al., 2017). The inbreeding trends across the genome are changed to genome-wide inbreeding estimates because of the introduction of whole-genome sequencing. The best example is of the organisms that contain homozygous regions because of the recombination that occurs with time, the inbreeding events, and deleterious fitness effects with the increase in homozygosity are quickly accessed (Hedrick and Garcia, 2016). The outcome of fitness across the specific variants of genes (sequences that code for proteins) is examined by comparing models of the protein with the conservative lines, which permits the estimation of the genetic load of deleterious alleles borne by the population species (Flanagan et al., 2010). When the inbreeding patterns are combined and compared with the fitness information across the genome, phenotypic defects caused in the homozygous regions due to inbreeding depression and deleterious alleles are quite visible (Kardos et al., 2016). Hike in the heterozygosity is a hub of recessive lethal alleles, whereas an individual consisting homozygous for the deleterious allele will not survive. This information is beneficial for the researcher who is working on the genetic rescue program.

The researchers can easily access genetic diversity decrease within an ecosystem and its degree of inbreeding. They can also predict the negative impact of inbreeding on the population's health and expected improvement due to the genetic rescue. A researcher can systematically plan all the things, which can locate the population and the individuals who are fit for the translocation into the population at risk. It should also be noted that editing at the genome level for allele replacement enhances the resistance to the diseases and adaptation to the changing climate. The best example is *Agrobacterium*-mediated transformation which is used to inject a wheat gene into the American chestnut tree (this tree is almost extinct due to the fungal pathogen). This genetically modified American chestnut strain starts resistance to the blight while the others succumb to the spectacle. This approach integrates the blight resistance to fungus, producing blight-resistant variety to conserve its diversity in the native place (Newhouse et al., 2014). Ecogenomics theoretically enable researchers to measure organisms' genetic diversity from prokaryotes to eukaryotes with due course of time (Tian et al., 2017). A most noticeable result of genomics is to clarify how much is shared within species (Xu, 2012). The diversity in the life form includes microorganisms, plants, insects, and their habitats/genes they produce (Khan et al., 2016).

Rao and Hodgkin in 2002 described biodiversity as the overall change in species, their genetic material, and the environments in which they live. The biodiversity is assessed by three regions, including species diversity, genetic diversity, and ecosystem diversity (Nuijten et al., 2016). In 2007, the convention on biological diversity stated that biodiversity conservation wholly depends on three pillars genetics, species, and ecosystem. If you have to stabilize, you need proper management in all of them. In 2018, Coates along with his research team concluded that species are not alike at the genetic level, but they differ, and it becomes the base for the conservation point of view. We can easily understand this by the phenomenon that the distantly related species exchange genes on hybridization. The variation at the gene level has taken place to withstand the harsh conditions in an environment. Many biotic/abiotic factors have a detrimental effect on biodiversity (Bahrndorff et al., 2016) like competition, parasitism, predation, disease, and separation due to human behavior, ecosystem alteration, natural disasters, and climate change (Khan et al., 2016), the invasive plant's introduction, natural ecosystems destruction, and natural components killing of the population (IUCN, 2015). These factors reduce plants and animals' biodiversity and cause the biodiversity of individual species to be extinct (De Vos et al., 2015). The elements are responsible for plant's expulsion from their natural environments and disappearing entirely from the wilderness (Mazzotti, 2014).

As a consequence of the risks mentioned above, several experimental studies have suggested that genomic tools' use proved an efficient strategy for conserving biodiversity in plants. The critical concern is to save endangered plants using a genomic approach (Aravanopoulos et al., 2015; Nuijten et al., 2016). Therefore, based on genomic research, genomic conservation methods have been established to avoid biodiversity loss. The most commonly genomic tools used for the conservation of species/populations include omics (genomics, proteomics, transcriptomics, and metabolomics). Some DNA markers, AFLP, SSR, RAPD, RFLP, and SSCP are also employed. These DNA methods are used to evaluate genetic diversity in species or populations (Khan et al., 2016). Due to these variations, an individual of a particular population can have an elite genotype that will allow it to withstand the desired conditions in an environment.

The more an individual is diversified, the more it will survive to produce its progeny with the desired allele for survival. McCormack et al., in 2013, suggested that if genomics needs to be used for conservation purposes, additional markers should be used on the genomes of the studied species sampled at significant distances.

In 2018, Grabowski and Rappsilber published a review in which they concluded the various omics used in integration to elucidate different functions for better understanding. This multiapproach is proliferating because their power of analysis increases when used in combination (Fig. 9.2). Combining genomics and transcriptomics helps find the genetic variations essential for various functions and diseases. Transcriptomics and proteomics help to study posttranslation mechanisms. In contrast, the synthesis, regulation, and degradation of proteins and relations at each level are studied using proteomics and metabolomics. Differentiation of stem cell and gene expression influenced by the genome are characterized by the epigenetic, transcriptomics, and proteomics combination. Last but not the least, phenotypic characters affected by the variation due to gene expression and genetic sequence are visualized by the three omics approaches, that is, phenomics, genomics, and transcriptomics. These

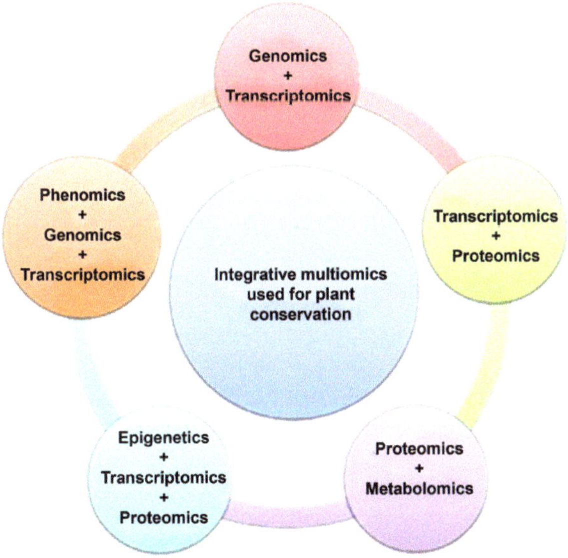

FIGURE 9.2 Integrative multiomics approach used for plant conservation.

integrations of various omics defend the challenges we face in ecogenomics for conservation and management.

9.7 CONCLUSIONS AND FUTURE DIRECTIONS

Advancement in genomics directly affects the study of ecogenomics. The biodiversity conservation and generation of new genetic data help the researchers to understand better relationships occurring in the ecosystem. The innovative approaches of genomics can check the plant diversity loss as it is crucial to know the genetic architecture, phylogenetic relationship, and loss of genetic variability in plants. Essential genes and genetic pathways are identified with the help of ecogenomics to study the functional variation in nature, which in turn characterizes the evolutionary and ecological ramification. Achieving this target would inevitably need a multidisciplinary approach, and using methods from various fields of biology in this research program is not so easy. The ecogenomics area is the most feasible because of the currently available genomic resources in extensive form. The development of transgenic plants is one of the opportunities to diversify the species in this era. Genomic techniques are presently being developed for various organisms with a rich history of ecological studies. These species are likely to work as model systems for ecogenomic research. Next-generation sequencing will promote advances in ecogenomic research, and it will allow genome sequencing in a cost-effective manner, leading to good research with effective results. In this way, we will start ecological genomics research in various plant species for their conservation and management. So the path to the top is challenging. Still, the perspectives into the genome work and the ecological, genetic mechanisms will be rewarding and supportive to humankind in conserving biodiversity.

ACKNOWLEDGMENTS

We sincerely thank Prof. Iqbal Parwez, Dean Academic Affairs and Dean, School of Biosciences and Biotechnology, BGSB University, Rajouri, for providing the necessary facilities to carry out this work. During this work, financial support from the Department of Biotechnology, Govt. of India, New Delhi, to Dr. Tanvir-Ul-Hassan Dar under grant no: BT/PR29155/FCB/125/6/2018 is highly acknowledged.

KEYWORDS

- **ecogenomics**
- **ecosystem**
- **biodiversity**
- **conservation**
- **management**
- **adaptation**

REFERENCES

Alonso, J. M.; Ecker, J. R. Moving forward in Reverse: Genetic Technologies to Enable Genome-Wide Phenomic Screens in Arabidopsis. *Nat. Rev. Genet.* 2006, *7*, 524–536.

Antonella, P. A.; Luca, G. The Quantitative Real-Time PCR Applications in the Monitoring of Marine Harmful Algal Bloom (HAB) Species. *Environ. Sci. Pollut. Res.* 2013, *20*, 6851–6862. https://doi.org/10.1007/s11356-012-1377-z

Aravanopoulos, F. A.; Ganopoulos, I.; Tsaftaris, A. Population and Conservation Genomics in Forest and Fruit Trees. *Adv. Bot. Res.* 2015, *74*, 125–155.

Bae, K. M.; Kwon, Y. S.; Cho, I. H.; Yi, S. I. Use of cDNA-AFLP for Transcript Profiling in Narrow Genetic Pools; For Example, Cucumber (*Cucumis sativus* L.). *Plant Breed.* 2006, *125*, 488–492.

Bahrndorff, S.; Alemu, T.; Alemneh, T.; Nielsen, J. L. The Microbiome of Animals: Implications for Conservation Biology. *Int. J. Genom.* 2016, 5304028. doi: 10.1155/2016/5304028

Baldwin, I. T.; Halitschke, R.; Paschold, A.; von Dahl, C. C.; Preston, C. A. Volatile Signaling in Plant–Plant Interactions: 'Talking Trees' in the Genomics Era. *Science* 2006, *311*, 812–815.

Breyne, P.; Dreesen, R.; Cannoot, B.; Rombaut, D.; Vandepoele, K.; Rombauts, S.; Vanderhaeghen, R.; Inzé, D.; Zabeau, M. Quantitative cDNA-AFLP Analysis for Genome-Wide Expression Studies. *Mol. Genet. Genom.* 2003, *269*, 173–179.

Breyne, P.; Zabeau, M. Genome-Wide Expression Analysis of Plant Cell Cycle Modulated Genes. *Curr. Opin. Plant Biol.* 2001, *4*, 136–142.

Brinke, A.; Reifferscheid, G.; Klein, R. et al. Transcriptional Changes Measured in Rice Roots after Exposure to Arsenite-Contaminated Sediments. *Environ. Sci. Pollut. Res.* 2018, *25*, 2707–2717. https://doi.org/10.1007/s11356-017-0515-z

Chariton, A. A.; Court, L. N.; Hartley, D. M.; Colloff, M. J.; Hardy, C. M. Ecological Assessment of Estuarine Sediments by Pyrosequencing Eukaryotic Ribosomal DNA. *Front. Ecol. Environ.* 2010, *8* (5), 233–238.

Cnudde, F.; Moretti, C.; Porceddu, A.; Pezzotti, N.; Gerats, T. Transcript Profiling on Developing Petunia Hybridafloral Organs. *Sex. Plant Reprod.* 2003, *16*, 77–85.

Coates, D. J.; Byrne, M.; Moritz, C. M. Genetic Diversity and Conservation Units: Dealing with the Species-Population Continuum in the Age of Genomics. *Front. Ecol. Evol.* 2018, *6* (165).

Colbert, T.; Till, B. J.; Tompa, R.; Reynolds, S.; Steine, M. N.; Yeung, A. T.; McCallum, C. M.; Greene, E. A.; Comai, L.; Henikoff, S. Throughtput Screening for Induced Point Mutations. *Plant Physiol.* 2001, *126*, 480–484.

Comai, L.; Young, K.; Till, B. J.; Reynolds, S. H.; Greene, E. A.; Codomo, C. A.; Enns, L. C.; Johnson, J. E.; Burtner, C.; Odden, A. R. Henikoff, S. Efficient Discovery of DNA Polymorphisms in Natural Populations by Ecotilling. *Plant J.* 2004, *37*, 778–786.

Convection on Biological Diversity, 2007. https://www.cbd.int/doc/meetings/cop-bureau/cop-bur-2007/cop-bur-2007-10-14-en.pdf

Crandall, K. A. et al. Considering Evolutionary Processes in Conservation Biology. *Trends Ecol. Evol.* 2000, *15*, 290–295.

Crow, J. F. The Odds of Losing at Genetic Roulette. *Nature* 1999, *397*, 293–294.

Curtis, T. P.; Sloan, W. T. Prokaryote Diversity and Its Limits: Microbial Community Structure in Nature and Implications for Microbial Ecology. *Curr. Opin. Microbiol.* 2004, *7*, 221–226.

Dallas, P. B.; Gottardo, N. G.; Firth, M. J.; Beesley, A. H.; Hoffman, K.; Terry, P. A.; Freitas, J. R.; Boag, J. M.; Cummings, A. J.; Kees, U. R. Gene Expression Levels Assessed by Oligonucleotide Microarray Analysis and Quantitative Real-Time PCR – How Well Do They Correlate? *BMC Genom.* 2005, *6*, Art. No. 59.

De Vos, J. M.; Joppa, L. N.; Gittleman, J. L. et al. Estimating the Normal Background Rate of Species Extinction. *Conserv. Biol.* 2015, *29* (2), 452–462.

Domsch, K. H. Effects of Pesticides and Heavy Metals on Biological Processes in Soil. *Plant Soil* 1984, *76*, 367–378.

Ebelechukwu, C.; Mmeka, A. A.; Kingsley, I.; Ubaoji; Arinze, B.; Nwokoye. Gene Silencing Technologies in Creating Resistance to Plant Diseases. *Int. J. Plant Breed. Genet.* 2014, *8*, 100–120.

Ekblom, R.; Wolf, J. B. W. A Field Guide to Whole-Genome Sequencing, Assembly and Annotation. *Evol. Appl.* 2014, *7*, 1026–1042.

Fay, J. C.; Wu, C. I. Sequence Divergence, Functional Constraint and Selection in Protein Evolution. *Annu. Rev. Genom. Hum. Genet.* 2003, *4*, 213–235.

Fire, A.; Xu, S. Q.; Montgomery, M. K.; Kostas, S. A.; Driver, S. E.; Mello, C. C. Potent and Specific Genetic Inference by Double-Stranded RNA in *Caenorhabditis elegans*. *Nature* 1998, *391*, 806–811.

Flanagan, S. E.; Patch, A. M.; Ellard, S. Using SIFT and PolyPhen to Predict loss-of-Function and Gain-of-Function Mutations. *Genet. Test Mol. Biomark.* 2010, *14*, 533–537.

Frankham, R. Genetic Rescue of Small Inbred Populations: Meta-Analysis Reveals Large and Consistent Benefits of Gene Flow. *Mol. Ecol.*, *24*, 2610–2618.

Frankham, R. Genetics and Conservation Biology. *C. R. Biol.* 2015, *326*, S22–S29.

Fujita, N.; Kazama, Y.; Yamagishi, N.; Watanabe, K.; Ando, S.; Tsuji, H.; Kawano, S.; Yoshikawa, N.; Komatsu, K. Development of the VIGS System in the Dioecious Plant Silenelatifolia. *Int. J. Mol. Sci.*, *20*, 1031.

Funk, W. C.; McKay, J. K.; Hohenlohe, P. A.; Allendorf, F. W. Harnessing Genomics for Delineating Conservation Units. *Trend. Ecol. Evol.* 2012, *27* (9), 489–496.

Gaggiotti, O. E. Genetic Threats to Population Persistence. *Ann. Zool. Fenn.* 2007, *40*, 155–168.

Gao, H.; Yang, Z. K.; Gentry, T. J.; Wu, L.; Schadt, C. W.; Zhou, J. Microarray-Based Analysis of Microbial Community RNAs by Whole-Community RNA Amplification. *Appl. Environ. Microbiol.* 2007, *73*, 563–571.

Grabowski, P.; Rappsilber, J. A. Primer on Data Analytics in Functional Genomics: How to Move from Data to Insight. *Trends in Biochemical Sciences* 2019, *44* (1).

Handelsman, J. Metagenomics: Applications of Genomics to Uncultured Microorganisms. *Microbiol. Mol. Biol. Rev.* 2004, *68*, 669–685.

Hansen, B. O.; Vaid, N.; Musialak, M.; Janowski, M.; Mutwil, M. Elucidating Gene Function and Function Evolution through Comparison of Co-expression Networks of Plants. *Front. Plant Sci.* 2014, *5*, 394. doi: 10.3389/fpls.2014.00394

Hedrick, P. W. Conservation Genetics: Where Are We Now? *Trends Ecol. Evol.* 2001, *16*, 629–636.

Hedrick, P. W.; Garcia-Dorado, A. Understanding in Breeding Depression, Purging, and Genetic Rescue. *Trends Ecol. Evol.* 2016, *31*, 940–952.

Hedrick, P. W.; Fredrickson, R. Genetic Rescue Guidelines with Examples from Mexican Wolves and Florida Panthers. *Conserv. Genet.* 2010, *11*, 615–626.

Heid, C. A.; Stevens, J. K.; Livak, K.; Williams, P. Real Time Quantitative PCR. *Genome Res.* 1996, *6*, 986–994. doi: 10.1101/gr.6.10.986

Held, M.; Gase, K.; Baldwin, I. T. Microarrays in Ecological Research: A Case Study of a cDNA Microarray for Plant-Herbivore Interactions. *BMC Ecol.* 2004, *4*: 13.

Hu, S. F.; Huang, Y. H.; Lin, C. P.; Liu, L. Y. D.; Hong, S. F.; Yang, C. Y.; Lo, H. F.; Tseng, T. Y.; Chen, W. Y.; Lin, S. S. Development of a Mild Viral Expression System for Gain-of-Function Study of Phytoplasma Effect or in Planta. *PLoS One* 2015, *10*, e0130139.

Hwang, H.; Yu, M.; Lai, E. Agrobacterium-Mediated Plant Transformation: Biology and Applications. *Arabidopsis Book* 2017, *15*, 20

IUCN. The IUCN Red List of Threatened Species 2015, 2017. http://www.iucnredlist.org/current-news (accessed Feb 22, 2017).

Jackson, R. B.; Linder, R.; Lynch, M.; Purugganan, M.; Somerville, S.; Thayer, S. S. Linking Molecular Insight and Ecological Research. *Trends Ecol. Evol.* 2002, *17*, 409–414.

Kadakkuzha, B. M.; Puthanveettil, S. V. Genomics and Proteomics in Solving Brain Complexity. *Mol. Bio Syst.* 2013, *9*, 1807–1821.

Kakirde, K. S.; Parsley, L. C.; Liles, M. R. Size Does Matter: Application-Driven Approaches for Soil Metagenomics. *Soil Biol. Biochem.* 2010, *42*, 1911–1923.

Kardos, M.; Akesson, M.; Fountain, T.; Flagstad, O.; Liberg, O.; Olason, P. et al. Genomic Consequences of Intensive Inbreeding in an Isolated Wolf Population. *Nat. Ecol. Evol.* 2017, *2*, 1–8.

Kardos, M.; Taylor, H. R.; Ellegren, H.; Luikart, G.; Allendorf, F. W. Genomics Advances the Study of Inbreeding Depression in the Wild. *Evol. Appl.* 2016, *9*, 1205–1218.

Keightley, P. D.; Eyre-Walker, A. Deleterious Mutations and the Evolution of Sex. *Science* 2000, *290*, 331–333.

Khaitovich, P. et al. (2006). Evolution of Primate Gene Expression. *Nat. Rev. Genet.* 2006, *7*, 693–702.

Khan, S.; Nabi, G.; Ullah, M. W. et al. Overview on the Role of Advance Genomics in Conservation Biology of Endangered Species. *Int. J. Genom.*, ID 3460416.

Kirkpatrick, M.; Jarne, P. The Effects of a Bottleneck on Inbreeding Depression and the Genetic Load. *Am. Nat.* 2000, *155*, 154–167.

Kondrashov, A. S. Contamination of the Genome by Very Slightly Deleterious Mutations – Why Have We Not Died 100 Times Over. *J. Theor. Biol.* 1995, 175, 583–594.

Kristensen, T. N. et al. Genome-Wide Analysis of Inbreeding Effects on Gene Expression in *Drosophila melanogaster*. *Genetics* 2005, *173*, 1329–1336.

Kyselková, M.; Kopecky, J.; Ságová-Marevcková, M.; Grundmann, G.; Moënne-Loccoz, Y. Oligonucleotide Microarray Methodology for Taxonomic and Functional Monitoring of Microbial Community Composition. *Plant Soil Environ.* 2009, *55* (9), 379–388.

Laikre, L. Hereditary Defects and Conservation Genetic Management of Captive Populations. *Zoo Biol.* 1999, *18*, 81–99.

Lande, R.; Shannon, S. The Role of Genetic Variation in Adaptation and Population Persistence in a Changing Environment. *Evolution* 1996, *50*, 434–437.

Lee, C. E.; Mitchell-Olds, T. Preface to the Special Issue: Ecological and Evolutionary Genomics of Populations in Nature. *Mol. Ecol.* 2006, *15*, 1193–1196.

Lee, S. H.; Walker, D. R.; Cregan, P. B.; Boerma, H. R. Comparison of Four Flow Cytometric SNP Detection Assays and Their Use in Plant Improvement. *Theor. Appl. Genet.* 2004, *110*, 167–174.

Loreau, M.; Naeem, S.; Inchausti, P.; Bengtsson, J.; Grime, J. P.; Hector, A.; Hooper, D. U.; Huston, M. A.; Raffaelli, D.; Schmid, B. et al. Biodiversity and Ecosystem Functioning: Current Knowledge and Future Challenges. *Science* 2001, *294*, 804–808.

Lyantagaye, S. L. Current Status and Future Perspectives of Bioinformatics in Tanzania. *Tanz. J. Sci.* 2013, 39.

Maphosa, F.; de Vos, W. M.; Smidt, H. Exploiting the Ecogenomics Toolbox for Environmental Diagnostics of Organohalide-Respiring Bacteria. *Trends Biotechnol.* 2010, *28*, 308–316.

Matsuyama, T.; Tamaoki, M.; Nakajima, N.; Aono, M.; Kubo, A.; Moriya, S.; Ichihara, T.; Suzuki, O.; Saji, H. CDNA Microarray Assessment for Ozone-Stressed *Arabidopsis thaliana. Environ. Pollut. Bark.* 2002, *117*, 191–194. 10.1016/S0269-7491(01)00320-7.

Mazzotti, F. J. The Value of Endangered Species: The Importance of Conserving Biological Diversity, Wildlife Ecology and Conservation Department, University of Florida, Cooperative Extension Service, Institute of Food and Agriculture Sciences, UF/IFAS Extension, Gainesville, FL, USA.

McCormack, J. E.; Hird, S. S.; Zellmer, A. J. et al. Applications of Next-Generation Sequencing to Phylogeography and Phylogenetics. *Mol. Phylog. Evol.* 2013, *66*, 526–538.

McKusick, V. A.; Ruddle, F. H. A New Discipline, a New Name, a New Journal. *Genomics* 1987, *1*, 1–2.

Meiklejohn, C. D.; Coolon, J. D.; Hartl, D. L.; Wittkopp, P. J. The Roles of Cis- and Trans-regulation in the Evolution of Regulatory Incompatibilities and Sexually Dimorphic Gene Expression. *Genome Res.* 2014, *24* (1), 84–95. doi:10.1101/gr.156414.113

Merila, J.; Crnokrak, P. Comparison of Genetic Differentiation at Marker Loci and Quantitative Traits. *J. Evol. Biol.* 2001, *14*, 892–903.

Newhouse, A. E. et al. Transgenic American Chestnuts Show Enhanced Blight Resistance and Transmit the Trait to T1 Progeny. *Plant Sci.* 2014, *228*, 88–97.

Nielsen, E. E. et al. Population Genomics of Marine Fishes: Identifying Adaptive Variation in Space and Time. *Mol. Ecol.* 2009, *18*, 3128–3150.
Nuijten, R. J. M.; Bosse, M.; Crooijmans, R. P. M. A. et al. The Use of Genomics in Conservation Management of the Endangered Visayan Warty Pig (Suscebifrons). *Int. J. of Genom.* 2016, ID 5613862.
Ohta, T. The Nearly Neutral Theory of Molecular Evolution. *Annu. Rev. Ecol. Syst.* 1992, *23*, 263–286.
Ostergaard, L.; Yanofsky, M. F. Establishing Gene Function by Mutagenesis in *Arabidopsis thaliana. Plant J.* 2004, *39* (5), 682–696. doi: 10.1111/j.1365-313X.2004.02149.x. PMID: 15315632.
Ouborg, N. J.; Vriezen, W. H. An Ecologist's Guide to Ecogenomics Soil DNA Libraries for Anticancer Drug Discovery. *J. Ecol.* 2007, *95*, 8–16.
Ozean, F.; Telci, C.; Kocak, N.; Yildiz, M.; Haspolat, I.; Tuna, E. Use of Genetically Modified Organisms in the Remediation of Soil and Water Resources. *Fresenius Environ. Bull.* 2012, *21*, 3443–3447.
Price, A. H. Believe It Or Not, QTLs Are Accurate! *Trends Plant Sci.* 2006, *11*, 213–216.
Rader, R. B. et al. Empirical Tests for Ecological Exchangeability. *Anim. Conserv.* 2005, *8*, 239–247.
Rao, V. R.; Hodgkin, T. Genetic Diversity and Conservation and Utilization of Plant Genetic Resources. *Plant Cell, Tissue Organ Culture* 2002, *68*, 1–19.
Reed, D. H.; Frankham, R. Correlation between Fitness and Genetic Diversity. *Conserv. Biol.* 2003, *17*, 230–237.
Reportlinker. Genomics Industry: Bioinformatics Market by Sector (Molecular Medicine, Agriculture, Research and Forensic), Segment (Sequencing Platforms, KnowledgeManagement Tools and Data Analysis Services) and Application (Genomics, Proteomics and Drug Design), Global Forecasts to 2017, 2013.
Reumers, J. et al. SNP Effect: A Database Mapping Molecular Phenotypic Effects of Human Non-Synonymous Coding SNPs. *Nucl. Acids Res.* 2005, *33*, D527–D532.
Robertson, D. VIGS Vectors for Gene Silencing: Many Targets, Many Tools. *Annu. Rev. Plant Biol.* 2004, 495–519.
Roy, K.; Tjard, D. C. B.; Joske, B. An Introduction to and a Reflection on the "Ecogenomics Promise". *J. Integr. Environ. Sci.* 2011, *8* (1), 23–38. doi: 10.1080/1943815X.2010.542165
Sanchita, Sharma, A. Gene Expression Analysis in Medicinal Plants under Abiotic Stress Conditions. *Plant Metabol. Regulat. Environ. Stress* 2018, *23*, 407–414. https://doi.org/10.1016/B978-0-12-812689-9.00023-6
Schena, M.; Shalon, D.; Davis, R.; Brown, P. Quantitative Monitoring of Gene Expression Patterns with a Complementary DNA Microarray. *Science* 1995, *270*, 467–470.
Schranz, E.; Manzaneda, A.; Windsor, A. J.; Clauss, M. J.; Mitchell-Olds, T. Ecological Genomics of Boecherastricta: Identification of a QTL Controlling the Allocation of Methionine- vs Branched-Chain Amino Acid-Derived Glucosinolates and Levels of Insect Herbivory. *Heredity* 2009, *102*, 465–474. doi: 10.1038/hdy.2009.12.
Siebert, P. D.; Chenchik, A.; Kellog, D. E.; Lukyanov, K. A.; Lukyanov, S. A. An Improved PCR Method for Walking in Uncloned Genomic DNA. *Nucl. Acids Res.* 1995, *23*, 1087–1088.

Snoeren, T. A. L.; de Jong, P. W.; Dicke, M. Ecogenomic Approach to the Role of Herbivore-Induced Plant Volatiles in Community Ecology. *J. Ecol.*. doi: 10.1111/j.0022-0477.2006.01183.x

Solanki, S.; Ameen, G.; Zhao, J. et al. Visualization of Spatial Gene Expression in Plants by Modified RNA Scope Fluorescent In Situ Hybridization. *Plant Methods* 2020, *16*, 71. https://doi.org/10.1186/s13007-020-00614-4

Sørensen, J. G. et al. Gene Expression Profile Analysis of Drosophila Melanogaster Selected for Resistance to Environmental Stress. *J. Evol. Biol.* 2007, *20*, 1624–1636.

Sousa, A. O. et al. Simulating Inbreeding Depression through the Mutation Accumulation Theory. *Phys. Act.* 2000, *278*, 563–570.

Spielman, D. et al. Most Species Are Not Driven to Extinction before Genetic Factors Impact Them. *Proc. Natl. Acad. Sci. USA* 2004, *101*, 15261–15264.

Straalen, N. M. V.; Roelofs, D. *An Introduction to Ecological Genomics*, 2nd ed.; Published in the United States by Oxford University Press Inc.: New York, 2012. ISBN 978-0-19-959468-9.

Sunyaev, S. et al. Prediction of Deleterious Human Alleles. *Hum. Mol. Genet.* 2001, *10*, 591–597.

Thomas, M. A.; Klaper, R. Genomics for the Ecological Toolbox. *Trends Ecol. Evol.* **2004,** *19*, 439–445.

Tian, W.; Paudel, D.; Vendrame, W.; Wang, J. Enriching Genomic Resources and Marker Development from Transcript Sequences of *Jatropha curcas* for Microgravity Studies. *Int. J. Genomics* 2017, 14.

Till, B. J.; Reynolds, S. H.; Greene, E. A.; Codomo, C.; Enns, L.; Johnson, J.; Burtner, C.; Odden, A.; Young, K.; Taylor, N.; Henikoff, J. G.; Comai, L.; Henikoff, S. Large-Scale Discovery of Induced Point Mutations with High-Throughput TILLING. *Genome Res.* 2003, 524–530.

Van Straalen, N. M.; Roelofs, D. *An Introduction to Ecological Genomics*, 1st ed.; Oxford University Press: Oxford, 2003.

Wang, X.; Tang, C.; Zhang, G. et al. cDNA-AFLP Analysis Reveals Differential Gene Expression in Compatible Interaction of Wheat Challenged with *Puccinia striiformis* f. sp. tritici. *BMC Genom.* 2009, *10*, 289. https://doi.org/10.1186/1471-2164-10-289

Waterhouse, P.; Wang, M. B. Gene Silencing as an Adaptive Defence against Viruses. *Nature* 2001, *411*, 834–842.

Woodworth, L. M. et al. Rapid Genetic Deterioration in Captive Populations: Causes and Conservation Implications. *Conserv.Genet.* 2002, *3*, 277–288.

Xiao, D. et al. cDNA-AFLP Analysis Reveals Differential Gene Expression in Incompatible Interaction between Infected Non-Heading Chinese Cabbage and *Hyaloperonospora parasitica. Horticult. Res.* 2016, *3*, 16034. doi:10.1038/hortres.2016.34

Xu, Y. Molecular Plant Breeding, in Molecular Breeding Tool: Omics and Arrays; CABI, 2012; p 68.

Young, N. D. QTL Mapping and Quantitative Disease Resistance in Plants. *Annu. Rev. Phytopathol.* 1996, *34*, 479–501. doi: 10.1146/annurev.phyto.34.1.479. PMID: 15012553.

CHAPTER 10

Genetic Diversity as a Conservation Tool

FAYAZ AHMAD DAR[1,2,*], INAYATULLAH TAHIR[3], and
REIAZ UL REHMAN[2]

[1]*Department of Bioresources, Amar Singh College,
Cluster University, Srinagar 190008, Jammu and Kashmir, India*

[2]*Department of Bioresources, University of Kashmir, Srinagar 190006,
Jammu and Kashmir, India*

[3]*Department of Botany, University of Kashmir, Srinagar 190006,
Jammu and Kashmir, India*

[*]*Corresponding author. E-mail: dfayaz5u@gmail.com*

ABSTRACT

Genetic diversity assessment is an indispensable tool that helps in prioritizing the conservation of both plant and animal species. It forms the basis of survival and adaptation, thus making it likely to acquire adaptive methods on which evolutionary success depends. The key objective of studying conservation biology is to maintain genetic variability at different levels and to act as a tool for population monitoring and analysis that assists in conservation planning. This genetic diversity is usually assessed within and among the individuals of a population/accessions/species at both phenotypic and genotypic levels. At the phenotypic level, genetic diversity is usually estimated with the help of quantitative and qualitative trait descriptors. However, at the genotypic level, it is evaluated with the

Plant Ecogenomics. Peerzada Arshid Shabir, PhD, Peerzada Yasir Yousuf, PhD &
Khalid Rehman Hakeem, PhD (Eds.)

help of biochemical markers, such as isozymes, allozymes, etc. as well as DNA-based molecular markers, such as restriction fragment length polymorphism (RFLP), amplified fragment length polymorphism (AFLP), inter-simple sequence repeat markers (ISSR), microsatellites or simple sequence repeats (SSR), and nowadays single nucleotide polymorphisms (SNPs) in the postgenomic era. The utilization of DNA-based molecular markers has revolutionized the field of conservation biology. This enables to measure the polymorphism, allelic frequency, diversity, population size, heterozygosity, genetic drift, etc. In this chapter, we discuss how this genetic variability is assessed with the help of various available tools and how effective they are in the conservation of important plant and animal species.

10.1 INTRODUCTION

Genetic diversity analysis at the intraspecific and interspecific levels allows better utilization of genetic resources in breeding and biodiversity conservation programmes. Consequently, information based on the genetic variation of existing germplasm plays an imperative role in its selection and potential development of cultivars. The importance of collection, characterization, evaluation, and management of plant genetic resources (PGRs) has been globally accepted as a prerequisite for the maintenance of agro-biodiversity. The collection of germplasm involves the identification of the sites, where the plant material is available in adequate quantity either in its wild or cultivated form. It is being characterized and evaluated based on the available attributes that can be easily assessed. Therefore, it helps in the classification of germplasm and development of core, mini-core, and reference sets based on universally accepted "descriptors" and "descriptor states." Furthermore, the systematic evaluation plays an important role in detailed analysis and documentation of germplasm depending on various morphological, physiological, and developmental traits studied under controlled as well as biotic and abiotic stress conditions. This enables the discovery and identification of new accessions, trait-specific accessions, and repatriated germplasm accessions conserved in genebanks of various countries across the globe. For an effective characterization and evaluation, the germplasm accessions are usually evaluated for a minimum of two consecutive years or vegetative phases whichever is earlier for the documentation and preparation of the crop catalog (Tripathi et al., 2018).

The characterization of germplasm involves the readings of highly heritable traits, generally expressed in all the environments. Germplasm, being an important reservoir of allelic diversity for traits that can be exploited in different breeding programmes irrespective of its conservation and management, is cost-intensive (Hintum et al., 2000). It should be initiated during the initial acquisition of germplasm and may be completed in a suitable environment preferably at a site nearest to the place of collection of the germplasm or under similar agro-climatic conditions. Evaluation of PGRs is of utmost importance to identify suitable germplasm with specific traits of interest and PGRs also act as an invaluable source of variation for future crop improvement programmes (Tripathi et al., 2018). Through a detailed evaluation of the germplasm, it gives an estimation of the critical value of genetic material, so that an appropriate conservation strategy can be framed. Moreover, evaluation of germplasm is a multidisciplinary approach as it involves collaborative effort from germplasm curator, plant breeder, physiologist, pathologist, entomologist, biochemist, and other experts. Based on morphological/phenotypic characterization of some important crop plants, it has become possible to effectively utilize the novel germplasm via the identification and selection of superior genotypes (Diederichsen and Richards, 2003). This facilitates the conservation of trait-specific genotypes/accessions that can be used in crop improvement by following conventional and non-conventional approaches.

Biodiversity is the repository that offers an opportunity to meet our food and nutritional requirements sustainable by maintaining natural genetic variation at intra- and interspecific levels in the biological systems. The main aim of conserving genetic variability is to improve the social and economic status of poor people to meet their food and nutritional requirements and at the same time keeping their cultural diversity intact. There is an unequal distribution of biological resources and thus the countries should collaborate for effective conservation and utilization at a global level (Rao and Hodgkin, 2002). Assessment of population structure and genetic variability in rare and endangered plant species is of vital importance for the development of conservation and management strategies because it provides valuable insights into the demography, reproduction, and ecology of species (Zaya et al., 2017).

10.2 FACTORS RESPONSIBLE FOR GENETIC DIVERSITY

The extent of variation observed within and among the individuals/genotypes/accessions of a population or species is defined as genetic diversity (Brown, 1983). It depends on some important factors which include the presence of different forms of a gene (alleles) among individuals, their interaction, distribution, and the influence they have on performance and total variation presented by the individuals of a population/species. This total variation is determined by the frequency of mutation and recombination. Other factors include selection, genetic drift, and the amount of gene flow in different populations that cause the variation in the population (Brown, 1988; Hamrick et al., 1992).

Furthermore, it is commonly believed that genetic diversity and structure in plant populations are determined by internal as well as external factors (Loveless and Hamrick, 1984). The quantification of the extent and distribution of genetic diversity in a population or species gives an idea about the population structure that plays a major role in its conservation. It has been generally observed that most of the conservation strategies whether *in situ* or *ex situ* have been devised with little knowledge about the genetic diversity of the species that is to be conserved. Therefore, for an effective conservation strategy, it is imperative to identify the key elements that are likely to affect genetics and allele distribution in a population. Several factors affect the genetics and allele distribution of a population, for instance, abiotic (edaphic, climatic, etc.) and biotic in addition to factors specific to populations like its selection and size, or the type of breeding system, ploidy, and genetic linkage of a species.

For the existence and adaptation of living organisms, analysis of genetic variability and population structure plays a key role in adaptive processes through the identification of novel genes that helps to overcome various biotic and abiotic stresses. It has been found that species with their narrow genetic base may not be able to adapt under fluctuating environmental conditions that tend to reduce fitness. Thus, for the successful long-term survival of a species, the presence of significant genetic variability is of fundamental importance for the efficient conservation and management of species (Rao and Hodgkin, 2002).

10.3 TOOLS USED TO ASSESS GENETIC DIVERSITY IN PLANTS

Previously, studies on genetic diversity in crop plants were exclusively based on the analysis of agro-morphological traits (Atta-Krah et al., 2004; Upadhyaya et al., 2010). Morphological/phenotypic markers involve the analysis of traits that can be visually characterized. The traits exhibit genetic polymorphism, which can be easily identified and manipulated. The plant material to be analyzed is selected and morphological trait descriptors like flower color, pigmentation, seed shape, seed size, growth habits, etc. are identified. Phenotypic markers have been found very useful in the formation of linkage maps and diversity analysis (Mondini et al., 2009). On the other hand, cytological markers involve the analysis of the structure and number of chromosomes, for instance, deletion, duplication, inversion, translocation, etc. The structural analysis of chromosomes are performed by chromosomal karyotyping and banding (Jiang, 2013). The advent of biochemical marker systems is closely associated with the improvements in biochemical and molecular biology tools (Lewontin and Hubby, 1966). The revolutionary technique of isozyme analysis has played a prominent role in molecular marker approaches. Isozymes and allozymes refer to the different molecular forms of an enzyme sharing a catalytic activity (Brown, 1978). These marker systems were more popular in the pre-genomic era. However, with the rapid evolution of new techniques, the development and utilization of molecular and biochemical markers have changed the way of analysis, identification, and exploration of PGRs and can be visualized as a significant breakthrough in the field of molecular biology and applied genetics. Nowadays, the availability of different types of DNA markers particularly single nucleotide polymorphisms (SNPs) has made it possible to use them for various analytical purposes, such as diversity studies, molecular taxonomy, phylogenetic relationships, conservation biology, mating systems, the level of inbreeding, fingerprinting, genetic mapping, cytogenetic mapping, physical mapping, diagnostics, plant variety protection and marker-assisted selection (MAS) in the postgenomic era. However, all these molecular techniques developed so far work on the principle to determine sequence polymorphism and repeat variations in the DNA/RNA as well as the mode of inheritance which also suggests that codominant marker systems tend to be more informative than dominant ones, besides, offering cheap, high-throughput, multidimensional genotyping for the analysis of PGRs. This has been largely possible due to unprecedented growth in the field of genomics resulting in a shift from arbitrarily primed DNA-based molecular markers to gene-targeted and functional markers (Gupta and Rustgi, 2004). With the development and expansion of public genomic databases and the exchange of sequence-related information, it has become possible to develop gene-targeted and functional markers, that are found in or near the genes of interest (Andersen and Lubberstedt, 2003). The availability of these markers has further improved the studies related to genetic variability, linkage map construction, and fingerprinting individuals or populations carrying useful genes (Kalendar et al., 2011). These markers have so many applications, such as selecting parental genotypes, linkage drag elimination during back-crossing, and selecting traits that are not possible through phenotypic characterization (Appleby et al., 2009). Therefore, the application of molecular techniques and statistical tools has helped in revealing genetic information about plant species/genotypes/accessions for their utilization in crop improvement programmes. Moreover, the wild relatives of crop plants considered to be the "reservoir of genes" also need to be characterized and evaluated for their effective utilization. Therefore phenotypic, biochemical, and molecular characterization and evaluation can play an important role in the management and utilization of PGRs for obtaining desirable results.

DNA sequencing involves the determination of four nucleotide bases, that is, A (adenine), G (guanine), C (cytosine), and T (thymine) present in the DNA molecule of particular plant species. This has become possible due to the rapid advancement in the field of sequencing technologies popularly known as next-generation sequencing (NGS) that has transformed the field of molecular biology and high-throughput sequencing techniques and is commonly used in phylogenetic, population genetic, and metagenomic studies (Margulies et al., 2005). First-generation sequencing also known as Sanger's sequencing was overtaken by second-generation sequencing, which is inexpensive and less time-consuming. The list of second-generation sequencing methods is massively parallel signature sequencing (MPSS), polony sequencing, Roche 454 pyrosequencing, illumina (Solexa) sequencing, SOLiD sequencing (Applied Biosystems), ion torrent semiconductor sequencing (New Life Technologies), PacBio (Pacific Biosciences), DNA nanoball sequencing and heliscope single-molecule sequencing

(Govindaraj et al., 2015). These sequencing methods are very efficient in conducting reliable population genetic studies and phylogenetic analysis of the whole genome. Presently, third-generation sequencing or long-read sequencing is under development as it can produce longer sequence reads as compared to second-generation sequencing (Christoph, 2016). This enables the sequencing of the whole genome in a cheaper, faster, and sophisticated way. Nanopore DNA sequencing and single-molecule real-time (SMRT) sequencing are the two examples of third-generation sequencing methods.

Markers have also been developed from chloroplast and mitochondrial DNAs obtained from plant genomes and have been found very useful in studying the genetic structure and phylogenetic relationships in plants. As both these DNAs show the uniparental mode of inheritance, their pattern of genetic differentiation is different compared with that of nuclear counterparts (Ahmed and Nielsen, 2020). Therefore, chloroplast and mitochondrial genomes provide information on variability analysis of plants in the same way as that of nuclear genomes. Besides, changes due to mutations in chloroplast and mitochondrial genomes are quite low than that of the nuclear ones as they are highly conserved in plants. Chloroplast DNA being maternally inherited is considered to be a powerful tool that can be extensively used in the studies of genetic variability and phylogenetics in plants (Ali et al., 1991; McCauley, 1994). In contrast, mitochondrial DNA is an unsuitable tool for genetic diversity and phylogenetic analysis. Consequently, based on the conserved sequences of chloroplast and mitochondrial DNA, new marker systems were developed.

Further, ribosomal RNA (rRNA) is employed to study the genetic variability at the nuclear level. These rRNA genes are available at specific chromosomal loci called *NOR* (nucleolus organizer region), arranged in tandem repeats and these repeated sequences are present in multiples in the genome. Many regions of these rRNA sequences are well conserved in eukaryotes, and therefore are suitable for phylogenetic analysis (Goffová and Fajkus, 2021). Besides, there are other regions in the genome, such as "Internal Transcribed Spacers (ITS)," markers based on such genomic regions are incorporated to differentiate populations intraspecifically (Fig. 10.1).

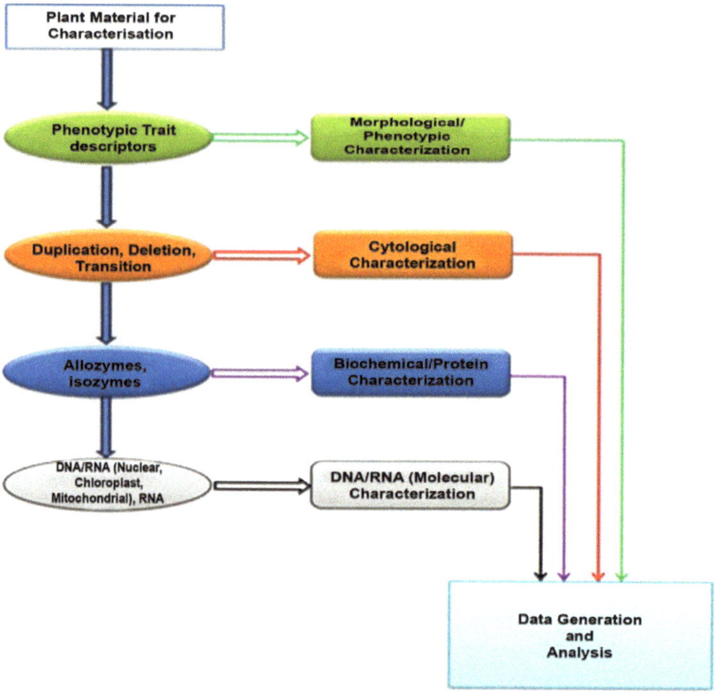

FIGURE 10.1 Different types of marker systems used for genetic diversity studies.

10.4 ANALYSIS OF GENETIC DIVERSITY AND SOFTWARE PROGRAMMES

The molecular data are calculated either manually or with the help of specific software is usually presented in binary form (0 and 1) for their absence/presence. Its analysis also depends upon the type of marker system used for a particular study and varies for dominant and co-dominant markers, due to the difference in banding patterns. Data obtained from genetic diversity studies need to be analyzed to draw meaningful inferences and subsequently their application. Two types of analysis are usually performed, first is to determine the genetic relationships and distances within and among the populations/genotypes and the second is to estimate the magnitude and distribution of genetic diversity. Therefore, for genetic relationships and distances, a matrix is constructed between samples/ genotypes pairwise. To calculate genetic distance and/or similarity among populations/genotypes/individuals, the various statistical measures used are Nei's and Li genetic diversity (Nei and Li, 1979), Jaccard's dissimilarity coefficient (Jaccard, 1908), simple matching coefficient (Sokal and Michener, 1958) and modified Roger's distance.

Furthermore, the resulting similarity or distance matrix is subjected to principal coordinate analysis, or through clustering, a dendrogram is constructed. PCA presents 2D or 3D scatter plot which is used to depict the genetic distances within and among the populations/genotypes. Similarly, a dendrogram or tree diagram represents similarity or dissimilarity among the populations/genotypes by forming clusters or groups of samples. There are different algorithms available for clustering, but the most extensively used ones are the unweighted pair group method with arithmetic mean (UPGMA), neighbor-joining (NJ) method, and Ward's minimum variance method (Karp et al., 1997).

Several software packages both paid and free versions are available for analyzing the data acquired during genetic diversity studies. Although most of these software programmes perform similar tasks, however, there are some differences in the user interface, type of data input and output, platform, and therefore its use entirely depends on the individual's preference and specific analysis needed. Some of the most commonly used software packages are Arlequin (Schneider et al., 2000; Excoffier et al., 2005), DnaSP (Librado and Rozas, 2009), PowerMarker (Liu and Muse, 2005), DARwin (Perrier and Jacquemoud-Collet, 2006), NTSYSpc (Rohlf, 2002), MEGA (Kumar et al., 2008; Tamura et al., 2011), PAUP (Swofford, 2002), STRUCTURE (Hubisz et al., 2009), POPGENE (Yeh Francis et al., 1999), GENEPOP (Raymond and Rousset, 1995), and GenAIEx (Peakall and Smouse, 2012).

10.5 IMPORTANCE OF STUDYING GENETIC DIVERSITY

For long-term success of populations or species, intraspecific genetic diversity is very important as it helps in revealing the adaptive potential of populations or species at the phenotypic level and also presents issues related to inbreeding and genetic drift that decreases the viability of populations due to neutral genetic variation (O'Grady et al., 2006; Hoffmann and Sgrò, 2011; Kahilainen et al., 2014). Therefore, decreased intraspecific genetic diversity either neutral or adaptive indicates an increased risk of extinction even in natural populations. The main aim of conservation approaches is to focus on the long-term endurance of species. The maintenance of intraspecific genetic diversity is of high significance. Although, there is an increasing need for intraspecific genetic diversity from the genetic and evolutionary point of view, contrarily its recognition in certain species-specific programmes (Ponce-Reyes et al., 2014) is important. It has also been observed that there is a considerable variation in data from species diversity thereby decreasing the importance of mapping intraspecific diversity (Walpole et al., 2009; Arponen, 2012). Therefore, it becomes an important criterion to assess the genetic diversity quite precisely at an intraspecific level that can be used to infer valuable information to be used in plant and animal-based genetic improvement programmes.

10.6 CONCLUSIONS AND FUTURE RECOMMENDATIONS

Genetic diversity analysis of diverse plant species is having pivotal importance as it helps in its conservation and management to meet the needs of the ever-growing human population. High genetic variability within and among the populations and genotypes can be employed in various crop improvement programmes as it provides options to develop new varieties and hybrids for the plant breeders. This has become possible as a result of morphological/phenotypic, biochemical, and molecular characterizations by making use of advanced tools and techniques like DNA and RNA-based marker systems. At times, the considerable size of genetic material may limit its use in breeding programmes that may be handled via the development of core, mini-core, and reference sets to represent the species collections exclusively. In recent times, the availability of cost-effective hardware, software, and significant progress in assay development and automation have made it possible to quantify the genetic diversity of plant populations and genotypes in less time with high accuracy. With the advent of second- and third-generation sequencing techniques, it has become possible to sequence the whole genome and generates significant data. Besides, there are many software packages to handle and analyze a large amount of data and yield appropriate results that can be utilized by the plant breeders to improve the results during crop improvement.

KEYWORDS

- **genetic diversity**
- **conservation biology**
- **population structure**
- **morphological markers**
- **molecular markers**

REFERENCES

Ahmed, N.; Nielsen, B. L. Plant Organelle DNA Maintenance. *Plants* **2020,** *9*, 683. doi:10.3390/plants9060683

Ali, I. F.; Neale, D. B.; Marshall, K. A. Chloroplast DNA Restriction Fragment Length Polymorphism in *Sequoia sempervirens* D. Don Endl., *Pseudotsuga menzipsii* (Mirb.) Franco, *Calocedrus decurrens* (Ton), and *Pinus taeda*. L. *Theor. Appl. Genet.* **1991,** *81*, 83–89.

Andersen, J. R.; Lubberstedt, T. Functional Markers in Plants. *Trends Plant Sci.* **2003,** *8*, 554–560.

Appleby, N.; Edwards, D.; Batley, J. New Technologies for Ultra-High Throughput Genotyping in Plants. In *Methods in Molecular Biology, Plant Genomics*; Gustafson, J. P., Langridge, P., Somers, D. J., Eds., Vol. 513; Humana Press: Totowa, NJ, 2009; pp 19–38.

Arponen, A. Prioritizing Species for Conservation Planning. *Biodiversity Conserv.* **2012,** *21*, 875–893.

Atta-Krah, K.; Kindt, R.; Skilton, J. N.; Amaral, W. Managing Biological and Genetic Diversity in Tropical Agroforestry. *Agroforestry Syst.* **2004,** *61*, 183–194.

Brown, A. H. D. Isozymes, Plant Populations Genetic Structure and Its Genetic Conservation. *Theor. Appl. Genet.* **1978,** *52*, 145–157.

Brown, A. H. D. The Genetic Diversity of Germplasm Collections. In *Proceedings of a Workshop on the Genetic Evaluation of Plant Genetic Resources*; Fraleigh, B., Ed.; Research Branch, Agriculture Canada: Toronto, Canada, 1988; pp 9–11.

Brown, W. L. Genetic Diversity and Genetic Vulnerability–An Appraisal. *Econ. Bot.* **1983,** *37* (1), 4–12.

Christoph B. Third Generation Sequencing: Technology and Its Potential Impact on Evolutionary Biodiversity Research. *Syst. Biodiversity* **2016,** *14* (1), 1–8.

Diederichsen, A.; Richards, K. Cultivated Flax and the Genus Linum L: Taxonomy and Germplasm Conservation. In *Flax: The Genus Linum*; Muir, A. D., Westcott, N. D., Eds.; CRC Press: London, 2003; pp 22–54.

Excoffier, L.; Laval, G.; Schneider, S. Arlequin (version 3.0): An Integrated Software Package for Population Genetics Data Analysis. *Evol. Bioinfo. Online* **2005,** *1*, 47–50.

Goffová, I.; Fajkus, J. The rDNA Loci—Intersections of Replication, Transcription, and Repair Pathways. *Int. J. Mol. Sci.* **2021,** *22*, 1302. https://doi.org/10.3390/ ijms22031302.
Govindaraj, M.; Vetriventhan, M.; Srinivasan, M. Importance of Genetic Diversity Assessment in Crop Plants and Its Recent Advances: An Overview of Its Analytical Perspectives. *Genet. Res. Int.* **2015,** *2015,* 431487–431487.
Gupta, P. K.; Rustgi, S. Molecular Markers from the Transcribed/Expressed Region of the Genome in Higher Plants. *Funct. Integr. Genom.* **2004,** *4*, 139–162.
Hamrick, J. L.; Godt, M. J. W.; Sherman-Broyles, S. L. Factors Influencing Levels of Genetic Diversity in Woody Plant Species. *New Forests* **1992,** *6*, 95–124.
Hintum, T. J. L.; Brown, A. H. D.; Spillane, C.; Hodgkin, T. Core Collections of Plant Genetic Resources. *IPGRI Techn. Bull.* **2000,** *3*, 48.
Hoffmann, A. A.; Sgrò, C. M. Climate Change and Evolutionary Adaptation. *Nature* **2011,** *470*, 479–485.
Hubisz, M. J.; Falush, D.; Stephens, M.; Pritchard, J. K. Inferring Weak Population Structure with the Assistance of Sample Group Information. *Mol. Ecol. Resour.* **2009,** *9* (5), 1322–1332.
Jaccard, P. Nouvelles Researches sur la Distribution Florale. *Bulletin de la Soci'et'e Vaudoise des Sciences Naturelles* **1908,** *44*, 223–270.
Jiang, G. L. Molecular Markers and Marker-Assisted Breeding in Plants. In *Plant Breeding from Laboratories to Fields*; Andersen, S. B., Ed.; Rijeka: *InTech*, 2013; pp. 45–83.
Kahilainen, A.; Keränen, I.; Kuitunen, K.; Kotioho, J. S.; Knott, K. E. Interspecific Interactions Influence Contrasting Spatial Genetic Structures in Two Closely Related Damselfly Species. *Mol. Ecol.* **2014,** *23*, 4976–4988.
Kalendar, R.; Flavell, A. J.; Ellis, T. H. N.; Sjakste, T.; Moisy, C.; Schulman, A. H. Analysis of Plant Diversity with Retrotransposon-Based Molecular Markers. *Heredity* **2011,** *106*, 520–530.
Karp, A.; Kresovich, S.; Bhat, K. V.; Ayad, W. G.; Hodgkin, T. Molecular Tools in Plant Genetic Resources Conservation: A Guide to the Technologies. *IPGRI Technical Bulletin 2, International Plant Genetic Resources Institute*, Rome, Italy.
Kumar, S.; Nei, M.; Dudley, J.; Tamura, K. MEGA: A Biologist-Centric Software for Evolutionary Analysis of DNA and Protein Sequences. *Brief. Bioinfo.* **2008,** *9* (4), 299–306.
Lewontin, R. C.; Hubby, J. L. A Molecular Approach to the Study of Genic Heterozygosity in Natural Populations. I. The Number of Alleles at Different Loci in *Drosophila pseudoobscura*. *Genetics* **1966,** *54*, 577–594.
Librado, P.; Rozas, J. DnaSP v5: A Software for Comprehensive Analysis of DNA Polymorphism Data. *Bioinformatics* **2009,** *25* (11), 1451–1452.
Liu, K.; Muse, S. V. PowerMarker: An Integrated Analysis Environment for Genetic Maker Analysis. *Bioinformatics* **2005,** *21* (9), 2128–2129.
Loveless, M. D.; Hamrick, J. L. Ecological Determinants of Genetic Structure in Plant Populations. *Annu. Rev. Ecol. Syst.* **1984,** *15*, 65–96.
Margulies, M.; Egholm, M.; Altman, W. E.; Attiya, S.; Bader, J. S.; Bemben, L. A.; Berka, J.; Braverman, M. S.; Chen, Y. J.; Chen, Z.; Dewell, S. B.; Du, L.; Fierro, J. M.; Gomes, X. V.; Godwin, B. C.; He, W.; Helgesen, S, He Ho, C.; Irzyk, G. P.; Jando, S. C.; Alenquer, M. L. I.; Jarvie, T. P.; Jirage, K. B.; Kim, J. B.; Knight, J. R.; Lanza, J. R.; Leamon, J. H.; Lefkowitz, S. M.; Lei, M.; Li, J.; Lohman, K. L.; Lu, H.; Makhijani, V. B.; McDade,

K. E.; McKenna, M. P.; Myers, E. W.; Nickerson, E.; Nobile, J. R.; Plant, R.; Puc, B. P.; Ronan, M. T.; Roth, G. T.; Sarkis, G. J.; Simons, J. F.; Simpson, J. W.; Srinivasan, M.; Tartaro, K. R.; Tomasz, A.; Vogt, K. A.; Volkmer, G. A.; Wang, S. H.; Wang, Y.; Weiner, M. P.; Yu, P.; Begley, R. F.; Rothberg, J. M. Genome Sequencing in Microfabricated High-Density Picolitre Reactors. *Nature* **2005,** *437*, 376–380.

McCauley, D. E. Contrasting the Distribution of Chloroplast DNA and Allozyme Polymorphisms among Local Populations of *Silene alba*: Implications for Studying Gene Flow in Plants. *Proc. Natl. Acad. Sci.* **1994,** *17*, 8127–8131.

Mondini, L.; Noorani, A.; Pagnotta, M. A. Assessing Plant Genetic Diversity by Molecular Tools. *Diversity* **2009,** *1*, 19–35.

Nei M.; Li, W. M. Mathematical Model for Studying Genetic Variation in Terms of Restriction Endonucleases. *Proc. Natl. Acad. Sci. USA* **1979,** *76* (10), 5269–5273.

O'Grady, J. J.; Brook, B. W.; Reed, D. H.; Ballou, J. D.;Tonkyn, D. W.; Frankham, R. Realistic Levels of Inbreeding Depression Strongly Affect Extinction Risk in Wild Populations. *Biol. Conserv.* **2006,** *133*, 42–51.

Peakall, R.; Smouse, P. E. GENALEX 6: Genetic Analysis in Excel. Population Genetic Software for Teaching and Research. *Mol. Ecol. Notes* **2012,** *6* (1), 288–295.

Perrier, X.; Jacquemoud-Collet, J. P. DARwin Software, 2006. http://darwin.cirad.fr/darwin

Ponce-Reyes, R.; Clegg, S. M.; Carvalho, S. B.; McDonald-Madden, E.; Possingham, H. P. Geographical Surrogates of Genetic Variation for Selecting Island Populations for Conservation. *Diversity Distrib.* **2014,** *20*, 640–651.

Rao, V. R.; Hodgkin, T. Genetic Diversity and Conservation and Utilization of Plant Genetic Resources. *Plant Cell, Tissue Organ Cult.* **2002,** *68*, 1–19.

Raymond, M.; Rousset, F. ENPOP: Population Genetics Software for Exact Tests and Ecumenicism. *J. Heredity* **1995,** *86*, 248–249.

Rohlf, F. J. *NTSYSpc: Numerical Taxonomy System, Version 2.1*; Exeter Publishing: Setauket, NY, USA, 2002.

Schneider, S.; Roessli, D.; Excoffier, L. *Arlequin: A Software for Population Genetics Data Analysis,* **2000,** *Version 2.0*; Genetics and Biometry Laboratory, Department of Anthropology, University of Geneva: Geneva, Switzerland, 2000.

Sokal, R. R.; Michener, C. D. A Statistical Method for Evaluating Systematic Relationships. *Univ. Kansas Sci. Bull.* **1958,** *38*, 1409–1438.

Swofford, D. L. *PAUP: Phylogenetic Analysis Using Parsimony (and Other Methods), Version 4*; Sinauer Associates: Sunderland, MA, USA, 2002.

Tamura, K.; Peterson, D.; Peterson, N.; Stecher, G.; Nei, M.; Kumar, S. MEGA5: Molecular Evolutionary Genetics Analysis Using Maximum Likelihood, Evolutionary Distance, and Maximum Parsimony Methods. *Mol. Biol. Evol.* **2011,** *28* (10), 2731–2739.

Tripathi, K.; Bhardwaj, R.; Bhalla, S.; Kaur, V.; Bansal, R.; Yadav, R.; Gangopadhyay, K. K.; Kumar, A.; Chaudhury, R. Plant Genetic Resources Evaluation: Principles and Procedures. *Indian Council of Agricultural Research-National Bureau of Plant Genetic Resources (ICAR-NBPGR)*, New Delhi., 2018

Upadhyaya, H. D.; Bramel, P. J.; Ortiz, R.; Singh, S. Geographical Pattern of Diversity for Morphological and Agronomic Traits in the Groundnut Germplasm Collection. *Euphytica* **2010,** *128*, 191–204.

Walpole, G. M.; Collen, B.; Collette, L.; Davidson, N. C.; Dulloo, E.; Fazel, A. M.; Galloway, J. N.; Gill, M.; Goverse, T.; Hockings, M.; Leaman, D. J.; Morgan, D. H.; Revenga, C.; Rickwood, C. J.; Schutyser, F.; Simons, S.; Stattersfield, A. J.; Tyrrell, T. D.; Zimsky, M. Ecology. Tracking Progress toward the 2010 Biodiversity Target and Beyond. *Science* **2009,** *325* (5947), 1503–1504. https://doi.org/10.1126/science.1175466

Yeh Francis, C.; Yang, R. C.; Boyle Timothy, B. J.; Ye, Z. H.; Mao Judy, X. POPGENE Version 1.32, the User-Friendly Shareware for Population Genetic Analysis. Molecular Biology and Biotechnology Centre, University of Alberta, Alberta, Canada, 1999. http://www.ualberta.ca/~fyeh/

Zaya, D. N.; Molano-Flores, B.; Feist, M. A.; Koontz, J. A.; Coons, J. Assessing Genetic Diversity for the USA Endemic Carnivorous Plant *Pinguicula ionantha* RK Godfrey (Lentibulariaceae). *Conserv. Genet.* **2017,** *18* (1), 171–180.

CHAPTER 11

Inbreeding and Genetic Load and Role of Ecogenomics in Landscape Genetic Restoration

MEENU GARG[1,+], PRAGYANSHREE NAYAK[1,+], ALOK KUMAR SINGH[2], and ASHWINI KUMAR RAY[1,3,*]

[1]*Department of Microbiology, Shaheed Rajguru College of Applied Sciences for Women, University of Delhi, Delhi 110096, India*

[2]*Department of Biochemistry, Shaheed Rajguru College of Applied Sciences for Women, University of Delhi, Delhi 110096, India*

[3]*Department of Environmental Studies, University of Delhi 110007, India*

[+]*Authors contributed equally.*

[*]*Corresponding author. Email: ashwinires21@gmail.com*

ABSTRACT

Taking into consideration the concept of "Inbreeding" which creates genetic load over various species and that leads to an urgent need for genetic restoration where "Ecogenomics" plays a key featuring role. The latest genre of Ecogenomics incorporates a few exploration headings. Among them, the developing field is landscape hereditary quality that is quickly growing into landscape genomic restoration. Landscape genetics analyzes the impacts of landscape and related variables based on the

Plant Ecogenomics. Peerzada Arshid Shabir, PhD, Peerzada Yasir Yousuf, PhD & Khalid Rehman Hakeem, PhD (Eds.)

hereditary design of populations. It emerged at the edge of landscape ecology and population genetic characteristics, and its subject matter is incredibly like that of populace genetic qualities. The study of the association between gene flows and indigenous variations and landscape highlights includes various determined procedures generally dependent on spatial data, including geostatistics, genetics, and landscape ecology. According to exploration, Ecological Genomics or Ecogenomics is a molecular approach in ecology that rectifies the morphological and functional properties of the genome for a better understanding of the interaction between organisms and their biotic and abiotic environment. The probability of getting offspring with two copies of recessive deleterious allele increases with an increase in inbreeding that gives rise to homozygosity. Reduction in fitness due to inbreeding depression leads to genetic load. Genetic load is the difference between the fitness of genotypes of mean and reference populations of different species. The variations observed can be of theoretical or optimal genotypic alleles. Genetic diversity within deme and cross-species of reintroduced population, for reintroduction geographical mapping of original material, are some of the highlights that are essential for the landscape genetic species to gain restoration. They are utilized to gauge the impacts of landscape factors based on genetic alteration and its investigation in landscape genetics needs both hereditary and topographical data. Ecogenomics in landscape restoration focuses on the recognition and separation of different genome-wide and loci-explicit impacts.

11.1 INTRODUCTION

Modern Plant Ecology had become a much more meticulous science in the late 19th century. Evolutionary concepts and genetic variations related to adaptation, alteration, and natural selection are the scientific pillars of Modern Ecological Theory. Its prime focus is on inbreeding depression, its causes, and drawbacks, how it is different from heterosis, genetic load, landscape, and ecological restoration, Ecogenomics (as a central role), its future challenges, and the leading ways to overcome them. Ecological genomics, the latest version of ecological genetics has emerged from the interaction of molecular and scientific methods that allow the assimilation of generally removed fields of science, like genomics and environment (Straalen and Roelofs, 2006). In the new version of environmental genetics, the emerging field is landscape genetics which is changing into landscape genomics. Landscape genetics came into being at the start of the 2000s with molecular population hereditary and landscape ecology (Manel et al., 2003). Like, population genetics, it also studies the interaction between the genetic structures of the population with the environmental conditions in which they inhabit. The major difference between population and landscape genetics is that landscape genetics functions in terms of individuals whereas population genetics takes into account whole population samples (Straalen and Roelofs, 2006). Ecological genomics, as a contemporary form of ecological genetics, covers a more extensive scope of issues and processes than its subclass, landscape genomics. Ecological genomics deals with the organism's structure and its interaction with the biotic and abiotic environments (Straalen and Roelofs, 2006). The chances of inbreeding depression certainly increase in small population size even in the random mating, as only a limited number of species contribute to each generation. The greater the extent of inbreeding depression more will be its contribution to the genetic load over species-to-species. Genetic load is the consequence of increased successive accumulations of deleterious genes. This can be decreased through a selection of heterozygotes, eliminating harmful mutations, and increased the frequency of population species (Crow, 1958). The accumulation of mutations and efficiency of elimination depends upon the rate of inbreeding and also the nature of mutations. Here, Ecological Restoration research extends its different levels of organization varying from gene to ecosystem. Genetic variations are considered for the successful implementation of restoration that can affect the choices from seed source selection to genetic control of an ecosystem. The major problem for restorationists is the utilization of local versus nonlocal plant material, as well as the mixture of source populations; preferably these choices are based on ecological and evolutionary theory research. Now only, Ecogenomics might be a possible boon for the restorationists as it provides a contemporary version of landscape genetics that serves as a discipline for an alliance of landscape genetics, population genetics, spatial statistics, and ecology. In turn, it inspects some microevolutionary processes that determine the genetic variations with species integration (Lee and Mitchell-Olds, 2006).

11.2 INBREEDING AND GENETIC LOAD

Inbreeding and genetic load are the two terms that go hand in hand as they are directly proportional to each other. Inbreeding is the crossing between related species that leads to the accumulation of recessive genes which in turn increases the genetic load over several populations. In a metapopulation, the rise to an age structure is due to the process of recurrent local extinction and recolonization among demes. Now a huge difference can be seen between the level and distribution of genetic diversity among the current subdivision of the population and previous ones. Slow progress in deme colonization was observed in the presence of mildly deleterious mutation due to an increase in within-species inbreeding depression and decline in genetic load with deme age.

11.2.1 INBREEDING DEPRESSION

Inbreeding prevails as cross-deme mating between relatives, or it may very well be simply the after effect of the population itself, that is, Inbred individuals within the same species are more related in contrast to the species as an entity (Lewontin, 1974).

Depression due to inbreeding is caused by increased individual homozygosity. It is predicted that inbreeding depression varies with fitness effect of involved deleterious allele, the mating species, population size, population genetic structure, condition of ambience to which they are exposed and duration of the individual affected (Crow, 1993). Descendants created by mating within individuals of similar species prompted more decrease in fitness as compared with that of parents from different species, a measure of heterosis. The inbreeding depression can be stated as the relative difference between the reductions in the fitness of inbred individuals to outbred individuals within similar species (Fig. 11.1).

11.2.2 GENETIC LOAD

Haldane and Muller initiated the concept of genetic load concerning mutation load. Mutation load is applicable throughout the genetic level as it is feasible to understand this notion via the perception of one gene locus. The offspring of species with low genetic load is more persistent than the species with an increased genetic load. High genetic load is responsible for the extinction of certain species as they have an accumulation of recessive alleles with homozygous characters. Severe species helped in purging the genetic load due to their lethal mutations and with a newly colonized population, there was a resultant decline in genetic load.

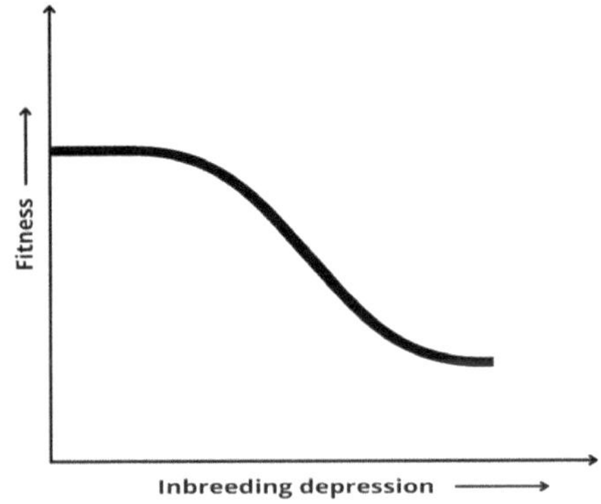

FIGURE 11.1 Graphical representation demonstrating inbreeding depression in self-fertile species.

11.3 ECOLOGICAL RESTORATION

Ecological restoration is the progression of support to reclaim deteriorated, demolished, or ruined ecosystems. It is an intended action that quickens an ecological trail—approaching an orientated condition. It is taken into consideration worldwide, as it is having negative influences on biological diversity and people's livelihoods.

11.3.1 NEED FOR ECOLOGICAL RESTORATION

Nowadays, it is realized that preservation and conservation of earth require restoration of ecology along with shielding acute regions. The four different elements that must be taken into consideration for successful ecosystem management are as follows:

Ecological restoration should:

(i) Improve conservation of biodiversity.

(ii) Expand Human livelihood.
(iii) Empower indigenous people.
(iv) To increase ecosystem efficiency.

This can be interpreted as restoration is considered as the chief component for sustaining and protecting the ecological biodiversity and ecosystem management.

11.3.2 PRINCIPLES OF GOOD ECOLOGICAL RESTORATION PRACTICE

It includes

- Sanctioning linkages in the broader landscape.
- Highlighting renovation process over organizational substitution.
- Allowing all shareholders especially marginalized resource users.
- Appealing all significant zones of society and authority, as well as the replaced and inadequate, in designing, enactment, and evaluation.
- Encompassing applicable collaborators in the description of borders for restoration.

A degraded ecosystem can be considered refurbished when it sustains sufficient living components to support its shape, processes, and functionality with negligible extrinsic help or funds. The modern strategy of the UN Convention on Biological Diversity (2011–2020) encircles the target of recovering 15% of the world's deteriorated ecosystems by 2020. The restoration of massive areas of wastelands that presently provide marginal economic or ecological value to landscapes generating goods and environmental services signifies an opportunity to increase our productive land area (Benayas and Bullock, 2012).

11.4 LANDSCAPE AND POPULATION GENETICS

The expression "Landscape Ecology" was conceived by K. Troll in 1939, on aerophotograph interpretation and pointed toward distinguishing the connections between the climate and vegetation. It came into light when the hypothesis of island biogeography was distributed by Mac Arthur and

Wilson in 1967. It clarified the variety of species of island fauna dependent upon the theory of a dynamical equilibrium between relocation from territory populaces and speculative extinction (Medinger et al., 2010). As of now, landscape genetics investigations are essentially centered on gene migration. They are utilized to survey the impacts of landscape factors on hereditary variety and elements. Nonetheless, it works on a more extensive geological and temporal scale and can uncover better subtleties of these processes than phylogeography (Schwartz and McKelvey, 2009). This order opens the route for all encasing causative examination of biodiversity. Landscape genetics depends on population genetics. S wright suggested that the gene migration assessment utilizes the hypothesis of isolation by distance in population genetics that shows, hereditary distances should be relative to the topographical distances.

The natural population system is distorted by several obstructions, such as mountain ranges, rivers, hills, valleys, etc. that in turn result in the migration of species to actively or passively least-cost paths which reflects the cost distances rather than geographical distances.

Landscape genetics is utilized to assess landscape connectivity, which uncovers the reliance between the activities of organisms and landscape association (Avise, 2008). The investigation of the relationship between gene streams and native varieties and landscape characteristics includes various determining procedures for the most part dependent on spatial information, including geostatistics, genetics, and landscape ecology. They are employed to assess the impacts of landscape factors on hereditary variety and its investigation in landscape hereditary qualities needs both genetics and topographical information. In modern times, Ecogenomics proves to be an effective tool for understanding various factors affecting ecology on a microscopic level. Ecological genomics deals with the working of the interaction of the genome of an organism and its biotic and abiotic components. Ecogenomics provide new direction to the researchers working on ecological restoration. The most recent genre of ecogenomics incorporates a few examination headings; among them, the developing aspect in landscape genetics that is quickly developing to landscape genomic restoration (Straalen and Roelofs, 2006). Landscape genetics analyzes the impacts of landscape and related natural variables on the hereditary construction of population (Storfer et al., 2007).

The main aim of ecogenomics is to illuminate the concept of adaptation. Plant ecology can easily adapt to the environment they grow in to sustain their needs, availability, and to balance natural selection processes with evolutionary mechanisms.

I. It aims to identify the possible reasons for phenotypic variations and genetic targets of natural selection. Using the ecogenomics tools, the evolving genes can be recognized along with their functionality (Lee and Mitchell-Olds, 2006).

Next, it explores certain new techniques for landscape and ecological restoration. It allows ecological processes and evolutionary forces to run parallel in hand (Jackson et al., 2002; Snoeren et al., 2007).

11.5 METHOD AND APPROACHES OF LANDSCAPE GENETICS AND GENOMICS

With time, various methods and approaches have been introduced and upgraded by researchers for expanding the study reach, to improve, and maximize the output. Among different types of molecular markers that are used for landscape genetics, microsatellites and amplified fragment length polymorphisms (AFLPs) give an appropriate resolution. Some of the neutral molecular markers are also employed in the studies of landscape genetics that deals with gene flow pattern (Tyutyunnik, 2004). The decision of genetic markers for landscape genetic investigations is moderately determined by a very important factor, that is, small size of landscapes. Due to the limited expansion or size of the landscape, polymorphic markers are used to study hereditary diversity (Holderegger, 2008). As indicated by an abridgment reviewed in 2010, it was discovered that microsatellites are associated with 70% and 31% of landscape genetic examinations in fauna and flora respectively (Storfer et al., 2010). Plant studies involve major utilization of allozymes (31%) and AFLPs (21%) (Storfer et al., 2010; Schwartz et al., 2010). As a limited understanding, it reduced to "promote the usage of genomic-sampled elements and to detach the loci-specific effects that include, conjugation, mating, and recombination from the genomic effects, such as genetic drift, genetic flow, and split for a better understanding of microevolution and its processes." Nonetheless, the utilization of genomic approaches empowers the recognizable proof of complex phenotypic qualities answerable for an organism adaptation. To study the genomics of plants, *Arabidopsis thaliana* proved to be the model species of plant as it has a small genome size and generation time. Moreover, an online calculation stage has been created inside the GENOME venture to encourage landscape genomics research. It involves numerous prevailing practices and creates a nonstop surface of hereditary differentiation with the concurrent evaluation of the impact created by various autonomous landscape factors (Landguth et al., 2010).

By 2008, the Mantel Test, Assignment Test, and some other tests based on distance model were widely used in analytical techniques of landscape genetics. The Mantel test is based on the statistical data that examines the relationship between matrices of genetic distances between population and subpopulation with that of geographical distances. However, the Assignment Test is based upon the usage of multilocus genotypic data that help in assigning an individual to a specific populace. These tests follow the fundamental standard of landscape genetics that proposes the need to work on the information of particular samples and not populace tests (Storfer et al., 2010).

The distinctive feature that is functional in landscape genetic investigations is a quantitative assessment of the impacts of landscape variations on gene flow and hereditary contrasts in populaces. The unadulterated populace does not utilize GIS information. Most strategies for landscape genetics are intended to assess the part of landscape highlights on the arrangement of connections among particular genotypes (Holderegger and Wagner, 2008)

Landscape ecological genetics and genomics are huge things in themselves, as their distribution among certain species of plants is quite extensive. The distinctive uses are on a broader spectrum (Manel et al., 2003).

11.5.1 SUBJECTIVE AND STRUCTURAL

E.B. Ford, in 1964, hypothesized that Ecological Genetics should consider the "changes and variations of the environment," consistent with this concept, the methodology should retain observations regarding experiment (Ouborg et al., 2010). In 1981, D. Merrell expressed that "Ecological Genetics is considered as the relation between inherent characteristics to those with the physical, biological, and environmental changes." The areas of Ecological Genetics were categorized in autecology and synecology. Synecology is a local area of complex populaces of several species forming one environment while the idea of autecology can be used in two ways. Firstly, it includes the standards of responses to environmental reactions on the organism level. Secondly, it analyzes the consequences of the environment on population species. A flowchart was made to define the structure of ecological genetics and its relation with exemplified heredity and variation **(Fig. 11.2)**.

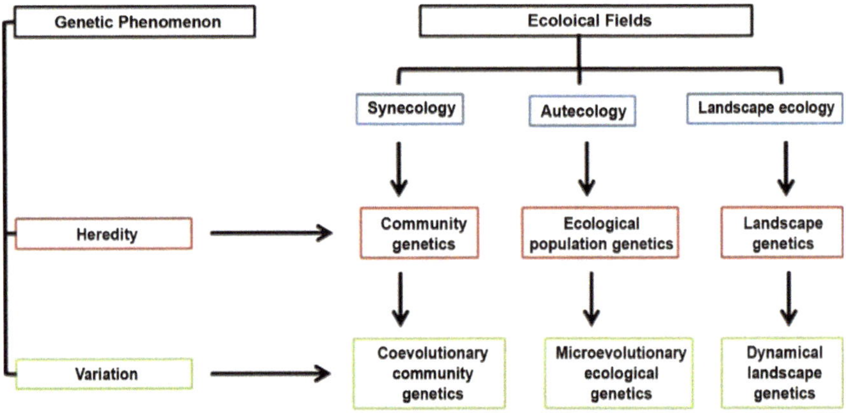

FIGURE 11.2 Basic structure of ecological genetics.

11.5.2 INSPECTION OF GENETIC APPLICATIONS IN RESTORATION MANAGEMENT

To examine the success of restoration projects just 42% of the studies were conducted that revealed the variation in genetic diversity pattern and the linkage of genetics with fitness demarcating underutilization in restoration projects. It was found that if foundation species possess high levels of homogeneity, then genetics models must be introduced that can alter the characteristics possessed by local communities. On the contrary, many empirical studies suggested that genetic guidelines must be based on life-history features (i.e., mating system) as it was assumed that species-explicit characteristics diminish the use of general measures of variations in a population. Such as, a nominal grassland herb (*Geranium pratense*) demonstrated quite less degree of genetic variability within the population. Conversely, *Alexgeorgea nitens* (a bisexual, clonal, and perpetual species) showed very undeniable degrees of genetic changes within a population (Ruiz-Jaen and Aide, 2005).

11.6 CHALLENGES AND OPPORTUNITIES IN LANDSCAPE GENETIC RESTORATION

It is claimed that genetic tools must be considered as the foundation of planning, estimation, evaluation, monitoring, and implementation of restoration projects. It is predicted that perceptual changes, advancement, and involvement of molecular techniques enable the use of hereditary information in recognized restoration programs that are currently underutilized. Genetic methods are complementary to the ecological methods that might be sinking and need immediate actions for their restoration to withstand the reproducing populations and willingly assist fitness parameters (Knapp and Dyer, 1998). Keeping in view, it is described that research directions and certain genetic approaches can supplement the vision of restoration in terms of community and functional ecosystems. Genetic studies indicate variations in species and affect ecosystem processes and the structure of communities' alignment, such as there was a huge diversity observed in *Solidago altissima*, the leading annual plant species, varies in community organization, arthropod diversity, and environmental adaptations (Craig, 2016). There are tremendous surveys and studies carried out by practitioners to highlight the fact that research and development programs need collaborations hand in hand. The advantage of these collaborations is the subsidization of professional and fundamental skills to facilitate the progression of infrastructure that could have significant profits. As we know, biodiversity is limited in the tropical regions, so studies on restoration genetics should be conducted to enhance the rate of overall restoration results and development programs (Trakhtenbrot et al., 2005; Jongejans et al., 2008; Damschen et al., 2014; Lohmus et al., 2014).

The practice of molecular data examining genetic variations is quite useful but it seems to be underutilized for the past few years. Several molecular and analytical techniques are emerging to diagnose available genetic methods. Among all the latest techniques that came into action, "Next Generation Sequencing (NGS)" requires special attention as this technique can elude all the shortcomings of molecular approaches. However, it is used to assimilate genetic variations among different species with the recognition of detrimental and adaptive genes as well as it determines the possibility through which genetic diversity is affected that can later improvise the decision-making processes. Meantime, NGS is looking forward to many pre-restoration mechanisms for the successful venture throughout faunal populations. It surveys species-richness in terms of metagenomics, epigenomics, and metabarcoding to evaluate short stretches of DNA present in the environment that can help in the re-establishment of natural resources and preservative methods for the benefit of the ecosystem (Gibson, 2002; Thomas and Klaper, 2004; Seifert et al., 2005; Ranz and Machado, 2006; Straalen and Roelofs, 2006).

Though the field of ecogenomics has proved to be of utmost significance, it also has certain limitations and faces numerous challenges throughout its practice.

- The first limitation is to expand this approach to every new species introduced in the natural environment, its habitat, ecological conditions, and life expectancy rates.
- The leading shortcoming faced by ecogenomics is to merge the fields of molecular biology with ecology and evolutionary biology as they needs to fine-tune the experimental designing for the integration of information regarding all the fields (Lesica et al., 2002).
- The next limitation lies in the allocation of data using high-throughput tools, such as microarrays, satellites, and radiolabeling that assist in the expression of genes for each faunal species. There are high chances of error sequences, so methodological data comprising alterations must be taken for the culture processes.
- This approach demands new research methods that are suitable for ecologists to outreach their attitude toward authorization and commandments in this field.

Various decisions have been made that can be implemented to overcome these certain issues. Amplifying restoration success needs an attentive decision-making process with consistent and precise information. Genetic tools maintain the nobility of this information through certain elements, limitations, and capabilities. As per the reports, 78% of the evaluation processes concentrated on the relation between the fitness of species and genetics. It was decided to choose the most suitable donor population to avoid species with increased outcrossing rates, and high genetic diversity was promoted. It is important to capture >95% of genetic variations in the species for a suitable sampling strategy (Leimu and Fischer, 2008). For the enhancement of restoration, gene flow either directly or indirectly must be stipulated to gather a broad array of information while maintaining local adaptation of the natural population.

11.7 CONCLUSION

In the past few years, Landscape genomics proves to be the most developing field in terms of research and microevolutionary processes. Landscape genomics examines the special features of the landscape and linked environmental aspects based on the hereditary organization of inhabitants. It emerged at the edge of landscape ecology and population genetics, and its topic is enormously like that of population genetics. During its research, the analysts felt several limitations, such as approaching and assessing every new species introduced in the natural environment, allocation of data using high-throughput tools, such as microarrays, satellites, and radiolabeling in culture processing that needs to be overcome. With time, novel methodologies that may refine research frameworks for both conservation genetics and evolutionary biodiversity have been introduced, such as SSRs, AFLPs, genetic surfaces, and molecular markers. It is recommended that ecological genetics must be considered from its budding stage to avoid misconception ideologies and to obtain identified restoration targets that may analyze ecological restoration outputs. Ecogenomics, as per the current scenario, is more oriented toward restoration practices so that it may reduce genetic load over species that may become extinct at a certain point in time.

The main cause of genetic load, that is, inbreeding is diminishing with the introduction of the ecogenomics approach and ecological restoration practices. It is raising high expectations to reverse the loss of ecosystem services and biological phenomena, including faunal species. If ecogenomics can satisfy all the expectations, it must encircle all the essential components that are required for the holistic view of ecological restoration advancement. Conceptual ideologies regarding conservation and intervention ecology must be equally important for improving ecological restoration. The alliance between restoration and genetics is gaining popularity these days as it can enhance the decision-making processes and is subtle for evolutionary studies.

KEYWORDS

- **ecogenomics**
- **inbreeding**
- **genetic load**
- **genotype**
- **landscape**

REFERENCES

Avise, J. C. Phylogeography: Retrospect and Prospect. *J. Biogeogr.* **2008,** *36* (1), 3–15.

Benayas, J. M. R.; Bullock, J. M. Restoration of Biodiversity and Ecosystem Services on Agricultural Land. *Mol. Ecol.* **2012,** *15*, 883–899.

Craig, T. P. Geographic Mosaic of Coevolution. In *Encyclopedia of Evolutionary Biology*; Kliman, R. M., Ed.; Academic Press, 2016; pp 201–207. ISBN 978012800**4265**. https://doi.org/10.1016/B978-0-12-800049-6.00194-3.

Crow, J. F. Some Possibilities for Measuring Selection Intensities in Man. *Hum. Biol.* **1958,** *30* (1), 1–13.

Crow, J. F. Mutation, Mean Fitness, and Genetic Load. *Oxford Surv. Evol. Biol.* **1993,** *9*, 3–42.

Damschen, E. I.; Baker, D. V.; Bohrer, G.; Nathan, R.; Orrock, J. L.; Turner, J. R.; Brudvig, L. A.; Haddad, N. M.; Levey, D. J.; Tewksbury, J. J. How Fragmentation and Corridors Affect Wind Dynamics and Seed Dispersal in Open Habitats. *Proc. Natl. Acad. Sci.* **2014,** *111* (9), 3484–3489.

Gibson, G. Microarrays in Ecology and Evolution: A Preview. *Mol. Ecol.* **2002,** *11*, 17–24.

Holderegger, R.; Wagner, H. H. Landscape Genetics. *BioScience* **2008,** *58* (3), 199–207.

Jackson, R. B.; Linder, R.; Lynch, M.; Purugganan, M.; Somerville, S.; Thayer, S. S. Linking Molecular Insight and Ecological Research. *Trends Ecol. Evol.* **2002,** *17*, 409–414.

Jongejans, E.; Skarpaas, O.; Shea, K. Dispersal, Demography and Spatial Population Models for Conservation and Control Management Perspect. *Plant Ecol. Evol.* **2008,** *9*, 153–170.

Knapp, E. E.; Dyer, A. R. When Do Genetic Considerations Require Special Approaches to Ecological Restoration? *Conserv. Biol.* **1998,** *2*, 345–363.

Landguth, E. L.; Cushman, S. A.; Schwartz, M. K.; Mckelvey, K. S.; Murphy, M.; Luikart G. Quantifying the Lag Time to Detect Barriers in Landscape Genetics. *Mol. Ecol.* **2010,** *19*, 4179–4191.

Lee, C. E.; Mitchell-Olds, T. Preface to the Special Issue: Ecological and Evolutionary Genomics of Populations in Nature. *Mol. Ecol.* **2006,** *15*, 1193–1196.

Leimu, R.; Kloss L.; Fischer, M. Effects of Experimental Inbreeding on Herbivore Resistance and Plant Fitness: The Role of History of Inbreeding, Herbivory and Abiotic Factors. *Ecol. Lett.* **2008,** *11* (10), 1101–1110.

Lewontin, R. C. *The Genetic Basis of Evolutionary Change*; Columbia University Press: New York and London, 1974.

Lohmus, K.; Paal, T.; Liira, J. Long-Term Colonization Ecology of Forest-Dwelling Species in a Fragmented Rural Landscape—Dispersal Versus. *Ecol. Evol.* **2014,** *4*, 3113–3126.

Manel, S.; Joost, S.; Epperson, B. K. Perspectives on the Use of Landscape Genetics to Detect Genetic Adaptive Variation in the Field. *Mol. Ecol.* **2010,** *19*, 3760–3772.

Manel, S.; Schwartz, M. K.; Luikart, G.; Taberlet, P. Landscape Genetics: Combining Landscape Ecology and Population Genetics. *Trends Ecol. Evol.* **2003,** *18* (4), 189–197.

Medinger, R.; Nolte, V.; Pandey, R. V.; Jost, S.; Ottenwälder, B.; Schlötterer, C.; Boenigk, J. Diversity in a Hidden World: Potential and Limitation of Next-Generation Sequencing for Surveys of Molecular Diversity of Eukaryotic Microorganisms. *Mol. Ecol.* **2010,** *19* (1), 32–40.

Ouborg, N. J.; Pertoldi, C.; Loeschcke, V.; Bijlsma, R. K.; Hedrick, P. W. Conservation Genetics in Transition to Conservation Genomics. *Trends Genet.* **2010,** *26*, 177–187.

Lesica, P.; Fred, W.; Allendorf, F. W. Ecological Genetics and the Restoration of Plant Communities: Mix or Match? *Restor. Ecol.* **2002,** *7* (1), 42–50.

Ranz, J. M.; Machado, C. A. Uncovering Evolutionary Patterns of Gene Expression Using Microarrays. *Trends Ecol. Evol.* **2006,** *21*, 29–37.

Roelofs, D.; Aarts, M. G. M.; Schat H.; Straalen, V. Functional Ecological Genomics to Demonstrate General and Specific Responses to Abiotic Stress. *Funct. Ecol. 200, 1*, 1120–1132.

Ruiz-Jaen, M. C.; Aide, T. M. Restoration Success: How Is It Being Measured? *Restor. Ecol.* **2005,** *13*, 569–577.

Schwartz, M. K.; McKelvey, K. S. Why Sampling Scheme Matters: The Effect of Sampling Scheme on Landscape Genetic Results. *Conserv. Genet.* **2009,** *10* (2), 441–443.

Schwartz, M. K.; McKelvey, K. S.; Cushman, S. A.; Luikart G. Landscape Genomics: A Brief Perspective. In *Spatial Complexity, Informatics, and Wildlife Conservation.* Cushman, S. A.; Huettmann, F., Eds.; Springer: Tokyo, 2010. https://doi.org/10.1007/978-4-431-87771-4_9

Seifert, M.; Scherf, M.; Epple, A.; Werner, T. Multievidence Microarray Mining. *Trends Genet.* **2005,** *21*, 553–558.

Snoeren, T. A. L.; De-Jong, P. W.; Dicke, M. Ecogenomic Approach to the Role of Herbivore-Induced Plant Volatiles in Community Ecology. *J. Ecol.* **2007,** *95*, 17–26.

Storfer, A.; Murphy, M. A.; Evans, J. S. Putting the "Landscape" in Landscape Genetics. *Heredity* **2007,** *98*, 128–142.

Storfer, A.; Murphy, M. A.; Spear, S. F.; Holderegger, W.; Waits, L. P. Landscape Genetics: Where Are We Now? *Mol. Ecol.* **2010,** *19*, 3496–3514.

Thomas, M. A.; Klaper, R. Genomics for the Ecological Toolbox. *Trends Ecol. Evol.* **2004,** *19*, 439–445.

Trakhtenbrot, A.; Nathan, R.; Perry, G.; Richardson, D. M. The Importance of Long-Distance Dispersal in Biodiversity Conservation. *Diversity Distrib.* **2005,** *11*, 173–181.

Tyutyunnik, Yu. G. O proiskhozhdenii i pervonachal'nom znachenii slova 'landshafta' [On the Origins and Initial Meaning of the Word 'Landscape']. *Izvestiya Rossiiskoi Akademii Nauk: Seriya Geograficheskaya* **2004,** *4*, 116–122.

Van Straalen, N. M.; Roelofs, D. *An Introduction to Ecological Genomics*, 2nd ed.; Oxford University Press: Oxford, 2012.

CHAPTER 12

Plant Ecogenomics: Challenges and Opportunities

BIPIN MAURYA,[1,2] LAKEE SHARMA,[1] SABITRI KUMARI,[1] NIDHI RAI,[1] and SHASHI PANDEY-RAI[1*]

[1]*Department of Botany, Institute of Science, Banaras Hindu University, Varanasi 221005, Uttar Pradesh, India;*
E-mail: bipinmaurya841@gmail.com

[2]*Laboratory of Morphogenesis, Centre of Advanced Study in Botany, Institute of Science, Banaras Hindu University, Varanasi 221005, Uttar Pradesh, India*

[*]*Corresponding author. E-mail: Shashi.bhubotany@gmail.com*

ABSTRACT

Plant ecogenomics is an adopted branch of science that comprises two major fields of biology, such as plant molecular biology and plant ecology. This field of science conveys information regarding changes in plant phenotype due to the environment, by using molecular tools and focuses on those genes which are regulated in evolution. Ecogenomics is not a very ancient clade of the biological sciences, as compared with other branches. To decipher the influential impacts of changing ecosystems on the holistic genetic makeup of biological entities, ecogenomics seems to function as an umbrella descriptor. Thus, this field is recognized as an emerging area for a better understanding of links between molecular techniques and ecology-related genomic obstacles like mutations and epigenetic alterations (DNA methylation, acetylation, phosphorylation, ubiquitination, etc.).

Plant Ecogenomics. Peerzada Arshid Shabir, PhD, Peerzada Yasir Yousuf, PhD & Khalid Rehman Hakeem, PhD (Eds.)

There are several applications of plant ecogenomic approaches, among them, the improvement of the crop yields and development of new varieties of economically important plants has gained more attention. Several epigenomic approaches, such as small RNAs (sRNAs) and DNA methylation used for genome-wide profiling, were first discovered in plants. These epigenomic approaches not only transformed the plant ecogenomic research but also creates similar approaches in fungal and animal systems. In this context, many research findings exist that describe the impacts of biotic causes, mainly in the form of herbivory and insect attacks on the genetic level of the plant. Despite a plethora of opportunities and advantages, studies related to ecological genomics harbor seldom hindrances and limitations. Major obstacles are problems in the maintenance of large samples and sequencing of their genome, available literature are pertaining to only model plant species, such as *Arabidopsis thaliana, Medicago* sp. and a dearth of tools to analyze ecogenomics data. Thus, there should be more concern toward nonmodel but economically, taxonomically, and evolutionary important plants and inventions of more techniques and tools to justify the reliability of molecular data related to ecological behavior of tester plant species. This chapter aims to provide an overview of plant ecogenomics and upcoming challenges and opportunities.

12.1 INTRODUCTION

Molecular biology is the field of science that deals with the functioning of the genome at the level of subcellular organization. In recent years, there have been several approaches and techniques in molecular biology used by plant ecologists. Plant ecologists use various approaches from molecular biology, especially in plant population biology, and many marker techniques available so far, including AFLP (amplified fragment length polymorphism), IISSR (inter simple sequence repeat), and microsatellite-PCR, to find out the variability in coding and noncoding regions. Plant ecologists utilize the top-down approach by applying the metabolomics acquisition methods, such as GC-MS, HPLC, LC-MS, FT-ICR-MS at plant diversity- environment interaction level to study the biochemical plasticity and genetic diversity which results in metabolite plasticity. Microarrays, SAGE, and NGS are applied to the monomorphic population to quantify gene expression differences with dynamic surroundings. Reverse genetics (gene silencing, overexpression) are applied to manipulate allelic variation to categorize the gene function which is of great importance specifically for the confirmation of gene function after the identification of gene by QTL mapping (Anderson and Mitchell-Olds, 2011). The Plant ecogenomic deals with the impact of surrounding on the genomic constitution of organisms as well as to elucidate the interaction between environment and phenotype of model and nonmodel plant species, using mechanistic and genomic processes.

Ecological genomics is the field where the two levels of life, largest and smallest meet. The research study can be performed at different levels, such as proteins to nucleic acids to cells to organisms, and finally at the levels of populations to communities up to ecosystems to biospheres. Ecological genomics uses a genomic explanation at the molecular level to describe the relation and interaction between environment and phenotype. Ecogenomics study focuses on the newly formed polymorphic trait for an inquiry at the genome and ecological level (Baldwin et al., 2006; Schranz et al., 2009).

An ecogenomics outlook is vital to find the solution to the enduring questions of evolutionary biology (Timmermans et al., 2008). The ecogenomic approach can throw light on the phylogenetic processes and molecular mechanisms that direct phenotypic alterations, elucidate the omics basis of morphological plasticity, and affirm the genomic foundation of selection and the genetic architecture of adaptation within the dynamic and spatially miscellaneous environment. Therefore, ecogenomics serves as the initial point for a better understanding of the relationship between ecology-related molecular studies. Ecogenomics integrates the subcellular realm of molecular biology where the genome functioning is studied and ecology deals with the interaction among the biotic and abiotic components of the environment. Using genomic language to solve the mysteries of phenotypic and environmental interaction as the foundation of explanation, ecogenomics opens the black box between the mechanism behind the genetic architecture of adaptation, phenotypic variation, and genetic target selection correlating with the dynamic environmental constraints (Ouborg et al., 2007).

When the different interdisciplinary clades of ecology, such as ecotoxicology, ecophysiology, and evolutionary ecology merged with molecular technologies, this resulted in the emergence of a new area of science that is ecogenomics (Fig. 12.1) (Kloet et al., 2011). In ecogenomics, further developments are expected to be the beginning of successful integration of omics technologies, such as transcriptomics and metabolomics, with community ecology (Snoeren et al., 2007). By implementation of these technologies, we can conserve biodiversity, minimize environmental pollution, study bioprospecting, improve environmental health, enhance bioremediation, and produces a bio-based economy (Delong, 2002).

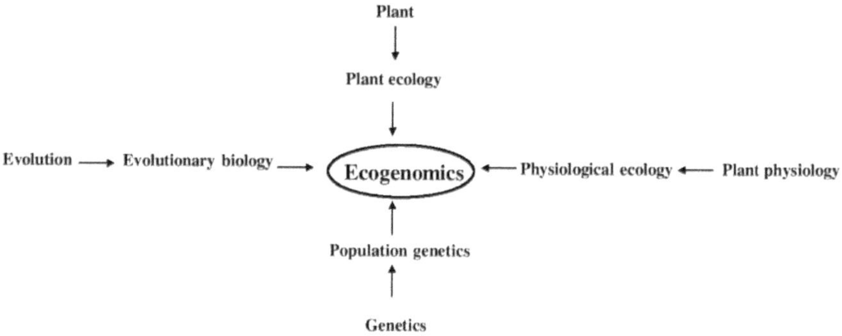

FIGURE 12.1 Emergence of ecogenomics by various fields of science.

Ecogenomics is easy to study in model organisms because ecogenomics tools are known in model species, for example, *Arabidopsis* and *Medicago sativa* or their close homologs in which case, many molecular studies are already performed. On behalf of these studies, data will also facilitate ecogenomics study in nonmodel species. Molecular techniques which are applied for a model organism can be used for closely related nonmodel species (Mitchell-olds, 2001). In this way, ecogenomics tools have put on the genetic basis of the plant' defensive characters in *Brassica, Boechera,* and other members of Brassicaceae, close relative to *Arabidopsis* (Broekgaarden et al., 2007; Schranz et al., 2009, Rushworth et al., 2011). Due to recent advancements in technology, ecogenomics tools are becoming worthwhile and appropriate for ecological investigation of nonmodel systems that are only far off related to model organisms.

Plants live in such an environment in which they interact with other organisms and these interactions are very diverse in which some are beneficial (plant–pollinator interactions) and some are not beneficial (herbivores or pathogens attacks). Consequently, plants have evolved a plethora of complex defense mechanisms to protect themselves against deleterious organisms. There is a complex mechanism by which plants can tolerate the biotic and abiotic stresses at the genome level (Tiwari et al., 2019) (Fig. 12.2). Plant responses are fine-tuned by interactions of signaling phytohormones (Nguyen et al., 2016). When more than one abiotic factor interacts with an individual plant, this may cause antagonistic or synergistic effects (Vescio et al., 2021) The apparent fields of ecogenomics and metabolomics will refine an understanding of the intricately regulated pathway that plants use to manage relationships with their biotic and abiotic environment (Van Dam, 2009). Ecogenomics assesses the type of defenses that evolved during temporal and spatial fluidity in herbivory and can advocate the similarity and dissimilarity in the genome of inducible and constitutive defenses. The ecogenomic study aims to understand the basis for adjustment and morphological alterations, investigate the mechanistic and evolutionary basis of species interactions, and finally to identify the affected gene. This field also detects the locus of genes on the chromosome responsible for ecologically important traits, examines the consequences under the natural condition of alleles at key quantitative trait loci (QTL), and also assesses how the abiotic and biotic factors alter the gene expression.

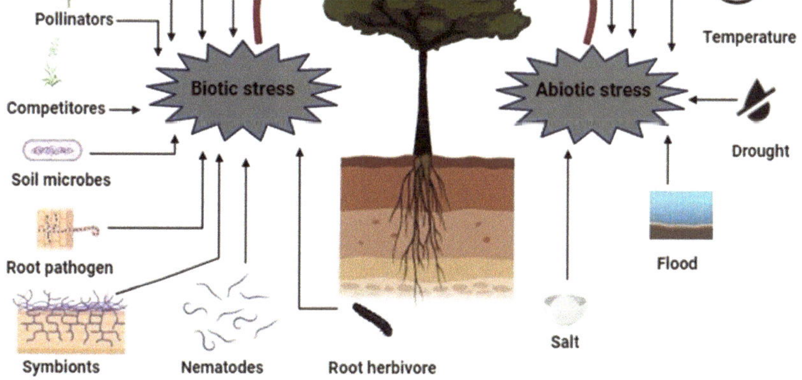

FIGURE 12.2 Diagrammatic representation of the interaction of biotic and abiotic factors to the plant.

12.2 ECOGENOMICS: APPROACHES AND OPPORTUNITIES

A very few complete sets of studies are available for few model species, such as *A. thaliana* and other nonmodel species (an adjacent relative of the model species), where the known tools might be applicable. This strategy is followed by many researchers focusing on phylogenetically close relatives of *Arabidopsis thaliana*, such as *A. lyrata* and *A. halleri* (Heidel et al., 2006; Filatov et al., 2006). Many genome-based schemes such as whole-genome sequencing projects provide the basis for the development of other tools and approaches to increase the opportunities in the field of ecogenomics, and are on their way to the increase the number of plant species. However, if one is going with a new plant species for which no particular genomic tools are available yet, we may face many limitations. So the hurdle with new species or nonmodel plants is the unavailability of ecogenomic approaches or the availability of very small information. Thus, an explorative and repetitive sequence program must be initiated, which may in most cases be unrealistic, costly, and time-consuming, because one is going to study the large genome. At the same time, retrieving and studying the complex sequence data will be very crucial in any ecogenomic program. This approach may be a bit haphazard for the ecologists because of the unavailability of sequence data (Long et al., 2019).

Thus, there can be two opportunities; one is to apply cDNA–AFLP to two well-defined samples that have different phenotypes to distinguish genes that are variably expressed in both samples. The retrieved differentially expressed ESTs (expressed sequence tags) can be delegated as a contender for the modulation of the morphological attributes under study and can be further accessed in several courses of action (Bohan et al., 2017). The cDNA microarray can be used to screen a larger number of samples. Hundreds of ESTs can be obtained from the PCR procedures and these can be sequenced relatively rapidly. Based on these retrieved EST data, one can examine the genotypic alterations responsible for the phenotypic changes (Fig. 12.3).

The second opportunity is to use previous knowledge from the model species. Due to the recent diversion of sequences between distantly located species, one cannot easily transfer the tools from one species to another (Seiler et al., 2019). Thus, the researcher must collect the information based on gene functions between species. Enough examples are showing the conserved nature of genes, such as the effect of sequence change in the *LEAFY* gene of nearly all flowering plant species on their flowering behavior in early-onset (Blázquez et al., 1997). Based on this, the degenerated PCR primers used to identify the species-specific version of the gene can further be used for species-specific RT-PCR detection.

Therefore, these techniques may be transferred from one species to another (Bomblies et al., 2003; Archambault and Bruneau, 2004). So, we can say that although the present tools are less favorable for the species of interest, various opportunities are also in hand to initiate a fruitful ecogenomics research program.

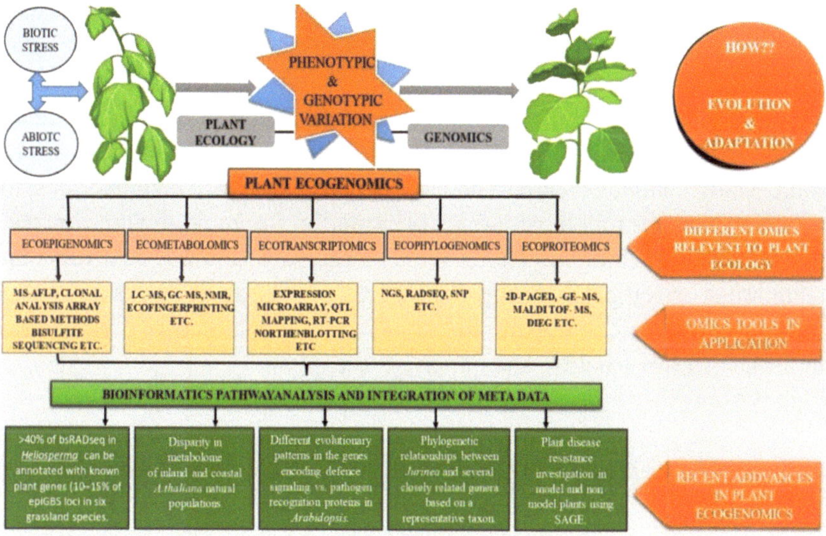

FIGURE 12.3 Different molecular techniques used in plant ecogenomics.

12.3 CHALLENGES OF ECOGENOMICS

The canopy of ecogenomic approaches confers the interdisciplinary associations between an individual's genomic infrastructure and holistic ecological impacts on it (Ungerer et al., 2008). Besides numerous applications and advantages of this field, certain challenges act as a remarkable hurdle toward the usefulness of this branch of science. The major challenges, often faced by ecologists and molecular biologists, during the amalgamation of their respective fields are greatly elaborated in this section.

Generally, ecologists pay more attention to identifying the mechanisms behind the structure of population, community, speciation, and adaptation. To fully understand this phenomenon, it becomes evident to comprehend the genetic basis (genome structure and function) to accomplish various omics technologies adopted by the ecologists. This has led to the addition of a new branch in science, that is, ecogenomics, which combines molecular sciences with ecology (Ungerer et al., 2008). Ecogenomics is a broad subject since it requires the use of various disciplines, such as biochemistry, bioinformatics, biotechnology, phylogenomic, proteomic, and transcriptomics, etc. Thus, in other words, exploitation of omic tools to unravel the ecological processes in plants is plant ecogenomics. Though plant ecogenomics is a promising field of science, there are certain developments required to be done. Ecology and molecular biology are two distinct science fields, and their research culture is varying in many parameters. Thus, combining both fields to give a fruitful outcome is not a smooth road. Many challenges are to be faced, such as for plant ecogenomics, the availability of complete genomics information for ecologically important species is not sufficient. There is also a scarcity of analytical tools and techniques specifically to dig the plant's ecogenomic data sets, and in addition to it, there is a requirement of research community collaboration in terms of infrastructure and expertise, etc.

12.3.1 NEED TO DEVELOP A MULTIDISCIPLINARY RESEARCH CULTURE

Ecogenomics is a powerful multidisciplinary branch that can answer the long-standing questions of evolution and adaptation. The problems and the questions may vary but certainly this field requires collaboration from the researcher community of all the relevant branches of sciences whether it would be bioinformatics, biochemistry, evolutionary biology, genomics, and physiology, etc. (Kloet et al., 2011). Ecologists and molecular biologists need to be aware of the genomics approach practiced in the nonmodel plant species by comparing them with thoroughly studied model plant populations that have shared some common evolutionary events during evolution. In a true sense, the evolution of the ecogenomics branch is possible if the international researcher community holds hands in expertise and research infrastructure availability (Ioana et al., 2015; Pawlowski et al., 2018).

12.3.2 INTEGRATION OF EXPERIMENTAL ECOLOGY AND EVOLUTIONARY BIOLOGY WITH MOLECULAR BIOLOGY

Ecology and molecular biology follow separate empirical procedures. Ecologists consider environmental disparity and genotypic variation in the study, and then accordingly design the experiment. In contrast, molecular biologists prefer to make the environment homogeneous with the least variation possible. An investigation was done on *Populus* sp. shows how both genomic and ecological experimental designs do not tune to give a complete ecogenomic picture (Liu et al., 2006). The effect of elevated levels of carbon dioxide (CO_2) on *Populus* ecophysical properties was studied with three levels of CO_2 concentration, three variable crosses, and around 324 biological replicates. The transcriptomic study on *Populus* for the effect of elevated levels of CO_2 was done with two levels of CO_2, one genetic clone, and a maximum of 12 biological replicates. The difficulty lies in linking both results as they show variability in magnitude therefore to incorporate both the subcellular realm and natural environment. It is required to design the experiments which integrate both field data in a complete sense (Taylor et al., 2005).

12.3.3 DEVELOPMENT OF SPECIFIC DATA ANALYTICAL TOOLS

Plant ecogenomics aims to study the mechanisms that govern the adaptation and evolution in plants with the help of omics tools (Kloet et al., 2011). The interaction of plants with the external stimulus in the dynamic environment holds a long evolutionary saga. The molecular tools can be beneficial in unraveling the hidden signaling networks responsible for the various biotic and abiotic interactions (Kaul et al., 2016). The genomics analysis is limited only to some plant species, such as *Arabidopsis, Medicago reticulata,* and some other crop species (Dhanapal et al., 2015). But with the advancement of omics tools like next-generation sequencing and microarray analysis, it is getting extended toward many ecologically important species also. For model plant species, the data acquired with advanced genomic tools like high throughput genome/transcriptome sequencing and microarray analysis can give thousands of gene expression for hundreds of individual species which can be difficult to study efficiently. This variation in the data is because of changing natural conditions, experimental errors, factorial inaccuracies, etc. So, it is necessary to develop analysis tools to study such a large set of ecogenomic data with more authenticity, for example, DAMBE7 software (Xia, 2018).

The study of significant evolutionary genes can be made through the transgenic approach by two methods. Firstly, by manipulating the allele on the polymorphic loci by overexpressing/silencing the gene, and secondly, through recreating the ancestral and derived genotype loci by reverse genetics. But in both cases, problems arise (Tian et al., 2003; Dean and Thornton, 2007). Gene silencing or overexpression in the complex biosynthetic pathway is supposed to duplicate the ancestral pattern or reflects the evolutionary lineage. Further, silencing of a gene may lead to a paralogous gene whose function is diverged from the gene of interest (Lustig, 2016). These issues need to be resolved explicitly. For the proper application of genomics technologies in ecology, there is a need to develop multivariate analysis tools that will lead to the suitable extraction of knowledge from the ever-increasing data sets (Van Dam and Poppy, 2008).

12.3.4 EXTENSION OF ECOGENOMICS ON NONMODEL PLANT SPECIES

There has been a lot of omics work done on the model and nonmodel crop species which aims to widen the understanding of subcellular functioning, for example, the evolution of plant herbivory defense strategy and their gene expression pattern exploration (Reymond et al., 2000; McKay and Stinchcombe, 2008; Luna et al., 2012). Molecular tools used in this context, for model plant species, can also be applied upon more closely related nonmodel species for ecogenomic studies (Mitchell-Olds, 2001). Ecogenomic tools in *Arabidopsis* are applied to study the genetic basis of plant defense strategies for its relatives *Brassica, Boechera,* and other members of the Brassicaceae (Broekgaarden et al., 2007). But for a detailed understanding of the evolution and existence of plant defense traits, it is necessary to apply ecogenomics on native species. The gene expression pattern among the same gene family members varies having the same coding sequences due to some regulatory changes (Okushima et al., 2005). These issues can be studied in model species. But the genomic data are not available for many ecologically important species that is why it is still a problem that must be resolved. Further, the integration of the genomics approach in nonmodel species becomes more difficult (Tagu et al., 2014). There are many additional reasons like large genome size and shallow knowledge about the genetic makeup. Polyploidy in plants makes genome sequencing intensive. Ecologists are applying transcriptomic data of model species and of related genera entirely to study nonmodel plant species, because of technical limitations (Tagu et al., 2014). Although, these explorations have an upper hand, since they are concentrated on a relatively insufficient number of species. But the problem they face is the availability of the genomic data conservative watchman species within the ecological field (Pinu et al., 2019). Though, this scenario will change with the advent of next-generation sequencing. So, we can contemplate this as a crucial era for ecological omics research (Sardans and Peñuelas, 2013).

12.4 RECENT ADVANCES IN PLANT ECOGENOMICS

Ecogenomics is the study of the effect of ecological constraints on the structure and function of genomes. This vast area of science is especially devoted to gain an understanding of the relationship between an organism and its environment. It not only deals with the relationship of the individual to its surroundings but also covers the way of adjustments and mechanisms involved during the interaction of the whole ecosystem at the community level (Schmitz, 2004). An understanding of these methods will empower the forecast to be made about the results of changes in the climate, landscape, and climate on populations and ecosystems. Ecology concerning different fields of genomics has directed us to enhance the ecogenomics study space. In the recent scenario, ecological genomic studies are not only limited to the core ecology and molecular biology, but it seems to give rise to many other such multidisciplinary branches of biological sciences (Anderson et al., 2011). Keeping these points in mind, this section deals with the recent advancements in the ecogenomic approaches, specifically focusing on the integration of different omics technologies into the ecological data.

12.4.1 ECOLOGICAL PROTEOMIC APPROACH

Environmental proteomics is steadily becoming a popular approach to study the ascending complex environments. The proteomic approach is used for the study of the evolutionary history of plants (Thiellement et al., 1999, 2002). With the advancement of increased genome and metagenome sequencing capabilities, the aid of quicker and more subtle mass spectrometers, management of large data sets has become possible. In this context, application of many proteomic technologies, such as native gel electrophoresis, 2-D PAGE (two-dimensional polyacrylamide gel electrophoresis), liquid chromatography techniques, isotope-coded affinity tag labeling, LTQ- Orbitrap MS (linear ion trap quadrupole orbitrap mass spectrometry), MALDI-TOF MS (matrix-assisted laser desorption/ionization–time of flight mass spectrometry), and SELDI (surface enhanced laser desorption/ionization) has been proven as a trouble shooter in solving the protein-related problems (Cho and William, 2007). Proceeding toward the next-generation protein studies, ecoproteomics is becoming a promising field to clear our understanding about the mechanisms behind competition strategy present in nature (Keller and Hettich 2009). Studies on *Arabidopsis* and rice have shown natural variation within and among different populations in varied environmental conditions through MALDI-TOF MS, suggesting that ecotype difference is a highly prevalent and convoluted phenomenon at the protein level (Thiellement et al., 1999; Diz et al., 2012).

12.4.2 ECOMETABOLOMIC APPROACH

Phenotypic responses of the plants are a metabolomic phenomenon that can be accessed by metabolomics. Metabolomics is an imperious means to study the phenotypic and genotypic variation in a particular genotype under different ecological circumstances specifically in stressful surroundings (Mittler and Shulaev, 2013). It is generally the application of metabolomics methods into environmental studies (Fiehn, 2002). The integration of ecogenomics into metabolomics can be beneficial in understanding the gene data networks involved in the adaptation process (Van Aggelen et al., 2010; Moshelion and Altman, 2015). It can add to our knowledge about the genes that govern phenotypic plasticity and evolutionary mechanism, genetic alteration in plants, and plant community composition (Hollert et al., 2007; Kuhlisch and Pohnert, 2015; Hong et al., 2016). The recent innovations in genomics tools, analytical, and *in silico* studies for large ecological data sets decoding and recoding have paved the way toward the understanding of the organism like health, population, adaptation, and defense mechanisms (Muller et al., 2018). The change in the metabolome profile under biotic and abiotic stress can be investigated through the ecometabolomic approaches, such as NMR (nuclear magnetic resonance), GC-MS (gas chromatography-mass spectrometry, and other data mining tools (Simmler et al., 2014; Wolfender et al., 2015). Ecometabolomic fingerprinting is a new approach toward the identification of the ecophysiologically significant metabolite to explain the different plant responses in an ecological context. There are many research reports on the use of the metabolomics approach to tackle ecological problems, such as herbivory, for example, in inland and coastal *Arabidopsis thaliana,* there is an inequality in the metabolite composition due to the variation in glucosinolate concentration (Kroymann, 2001). This has resulted in chemical-based adaptation in specialist and generalized herbivores (Arany et al., 2008).

12.4.3 EPIGENETIC APPROACH

Epigenetic changes are the outcomes of external changes and modification in DNA without modulating the genomic sequences. Most of the time, impacts of these alterations are visible in the species in the form of visual characters, that is, morphology of a particular plant (Des Roches et al., 2018). Intraspecific morphological differences are the outcome of their interaction with the surrounding and genetic variations. These phenotypic differences are governed by co-occurrences of the molecular variations through the epigenetic mechanism and environmental issues. Thus, the ecoepigenetic approach is a combination of ecological experimentation with epigenetic analysis to give a clear picture of plant adaptation in their habitat, phenotype, stress responses, and their distribution patterns. Latzel et al. (2016), in their report, has stated that epigenetic-based memory in clonal plants may be the basis of adopted behavior in the ecosystem.

12.4.4 ECOPHYLOGENOMIC APPROACH

Ecophylogenomics helps in understanding the evolutionary relationship and adaptation mechanism under dynamic environmental constraints. It specifically focuses on the ecologically significant gene segments. This can give an idea about the fitness status of the alleles at QTLs in nature. Also, it assesses the effect of biotic and abiotic environments on gene expression. It can also solve the phenomenon of genetic drift, speciation, and similar genetic pathways that are used by different taxons to achieve convergent phenotypic variation (Mitchell-Olds and Schmitt 2006; Mitchell-Olds et al., 2007). Recent ecogenomic techniques used are DNA barcoding, RAD seq, SNP, and NGS, etc. Current studies focussing on species discrimination with the help of DNA barcoding marker trnH-psbA have been a speedy and easy method (Pei et al., 2017). The Super matrix approach used in DNA barcoding could be used to reconstruct the community phylogenetic tree using three locus barcode combinations (rbcL + matK + psbA − trnH) (Tripathi et al., 2013). This will be useful in assigning an exact systematic position to flora and fauna. RAD seq study revealed that all the known mediterranean taxa are independent species excluding *Abies tazaotana* and *Abies marocana* (Sękiewicz, 2013). They have a complex ancient and current gene flow evolutionary history because of the interaction with the environmental factors.

ACKNOWLEDGMENT

The authors are thankful to CAS-Botany (BHU) and DST-FIST for providing funds to the Department of Botany, Institute of Sciences, Banaras Hindu University. Lakee Sharma and Sabitri Kumari are thankful to CSIR-New Delhi, India for providing financial support in the form of fellowship as SRF and JRF respectively.

KEYWORDS

- **plant ecogenomics**
- **epigenetics**
- **DNA methylation**
- **crop improvement**

REFERENCES

Anderson, J. T.; Mitchell-Olds, T. Ecological Genetics and Genomics of Plant Defences: Evidence and Approaches. *Funct. Ecol.* **2011,** *25* (2), 312–324.

Anderson, J. T.; Willis, J. H.; Mitchell-Olds, T. Evolutionary Genetics of Plant Adaptation. *Trends Genet.* **2011,** *27* (7), 258–266.

Arany, A. M.; De Jong, T. J.; Kim, H. K.; Van Dam, N. M.; Choi, Y. H.; Verpoorte, R.; Van der Meijden, E. Glucosinolates and Other Metabolites in the Leaves of *Arabidopsis thaliana* from Natural Populations and Their Effects on a Generalist and a Specialist Herbivore. *Chemoecology* **2008,** *18* (2), 65–71.

Archambault, A.;Bruneau, A. Phylogenetic Utility of the Leafy/Floricaula Gene in the Caesalpinioideae (Leguminosae): Gene Duplication and a Novel Insertion. *Syst. Bot.* **2004,** *29* (3), 609–626.

Baldwin, C.; Hienerth, C.; Von Hippel, E. How User Innovations Become Commercial Products: A Theoretical Investigation and Case Study. *Res. Pol.*, *35*(9), 1291-1313.

Blázquez, M. A.; Soowal, L. N.; Lee, I.; Weigel, D. Leafy Expression and Flower Initiation in Arabidopsis. *Development* **1997,** *124* (19), 3835–3844.

Bohan, D. A.; Vacher, C.; Tamaddoni-Nezhad, A.; Raybould, A.; Dumbrell, A. J.; Woodward, G. Next-Generation Global Biomonitoring: Large-Scale, Automated Reconstruction of Ecological Networks. *Trends Ecol. Evol.* **2017,** *32* (7), 477–487.

Bomblies, K.; Wang, R. L.; Ambrose, B. A.; Schmidt, R. J.; Meeley, R. B.;Doebley, J. Duplicate Floricaula/Leafy Homologs zfl1 and zfl2 Control Inflorescence Architecture and Flower Patterning in Maize. *Development*, *130* (11), 2385–2395.

Broekgaarden, C.; Poelman, E. H.; Steenhuis, G.; Voorrips, R. E.; Dicke, M.;Vosman, B. Genotypic Variation in Genome-Wide Transcription Profiles Induced by Insect Feeding: Brassica Oleracea–Pierisrapae Interactions. *BMC Genom.* **2007,** *8* (1), 239.

Cho William, C. S. Proteomics Technologies and Challenges. *Genom. Proteom. Bioinfo.* **2007,** *5* (2), 77–85.

Dean, A. M.; Thornton, J. W. Mechanistic Approaches to the Study of Evolution: The Functional Synthesis. *Nat. Rev. Genet.* **2007,** *8* (9), 675–688.

DeLong, E. F. Microbial Population Genomics and Ecology. *Curr. Opin. Microbiol.* **2002,** *5* (5), 520–524.

Des Roches, S.; Post, D. M.; Turley, N. E.; Bailey, J. K.; Hendry, A. P.; Kinnison, M. T.; Palkovacs, E. P. The Ecological Importance of Intraspecific Variation. *Nat. Ecol. Evol.* **2018,** *2* (1), 57–64.

Dhanapal, A. P.; Govindaraj, M. Unlimited Thirst for Genome Sequencing, Data Interpretation, and Database Usage in Genomic Era: The Road Towards Fast-Track Crop Plant Improvement. *Genet. Res. Int.* **2015.**

Diz, A. P.; Martínez-Fernández, M. O. N. I. C. A.; Rolán-Alvarez, E. M. I. L. I. O. Proteomics in Evolutionary Ecology: Linking the Genotype with the Phenotype. *Mol. Ecol.* **2012,** *21* (5), 1060–1080.

Fiehn, O. Metabolomics—The Link Between Genotypes and Phenotypes. In *Functional Genomics*. Springer: Dordrecht, 2002; pp 155–171.

Filatov, V.; Dowdle, J.; Smirnoff, N.; Ford-Lloyd, B. R. I. A. N.; Newbury, H. J.;Macnair, M. R. Comparison of Gene Expression in Segregating Families Identifies Genes and Genomic Regions Involved in a Novel Adaptation, Zinc Hyperaccumulation. *Mol. Ecol.* **2006,** *15* (10), 3045–3059.

Heidel, F.; Solem, F. K.; Breitenbuecher, F.; Lipka, D. B.; Kasper, S.; Thiede, M. H.; Feldman, E. Clinical Resistance to the Kinase Inhibitor PKC412 in Acute Myeloid

Leukemia by Mutation of Asn-676 in the FLT3 Tyrosine Kinase Domain. *Blood* **2006,** *107* (1), 293–300.
Hollert, H.; Seiler, T. B.; Blaha, L.; Young, A. L. Multiple Stressors for the Environment: Present and Future Challenges and Perspectives. *J. Soils Sediments* **2007,** *7* (4), 272.
Hong, J.; Yang, L.; Zhang, D.; Shi, J. Plant Metabolomics: An Indispensable System Biology Tool for Plant Science. *Int. J. Mol. Sci.* **2016,** *17* (6), 767.
Ioana, M. I.; Anda, M. I.; Ione, P.; Bianca, R. A. The Importance of Community Involvement in Research and Innovation between 2014–2020. *Procedia-Soc. Behav. Sci.* **2015,** *182,* 619–622.
Kaul, S.; Sharma, T.; Dhar, M. K. "Omics" Tools for Better Understanding the Plant–Endophyte Interactions. *Front. Plant Sci.* **2016,** *7,* 955.
Keller, M.; Hettich, R. Environmental Proteomics: A Paradigm Shift in Characterizing Microbial Activities at the Molecular Level. *Microbiol. Mol. Biol. Rev.* **2009,** *73* (1), 62–70.
Kloet, R.; de Cock Buning, T.; Bunders, J. An Introduction to and a Reflection on the "Ecogenomics Promise". *J. Integr. Environ. Sci.* **2011,** *8* (1), 23–38.
Kroymann, J.; Textor, S.; Tokuhisa, J. G.; Falk, K. L.; Bartram, S.; Gershenzon, J.; Mitchell-Olds, T. A Gene Controlling Variation in Arabidopsis Glucosinolate Composition Is Part of the Methionine Chain Elongation Pathway. *Plant Physiol.* **2001,** *127* (3), 1077–1088.
Kuhlisch, C.; Pohnert, G. Metabolomics in Chemical Ecology. *Nat. Prod. Rep.* **2015,** *32* (7), 937–955.
Latzel, V.; Rendina González, A. P.; Rosenthal, J. Epigenetic Memory as a Basis for Intelligent Behavior in Clonal Plants. *Front. Plant Sci.* **2016,** *7,* 1354.
Liu, N.; Dang, Q. L.; Parker, W. H. Genetic Variation of Populus tremuloides in Ecophysiological Responses to CO_2 Elevation. *Botany* **2006,** *84* (2), 294–302.
Long, G. S.; Hussen, M.; Dench, J.; Aris-Brosou, S. Identifying Genetic Determinants of Complex Phenotypes from Whole Genome Sequence Data. *BMC Genom.* **2019,** *20* (1), 1–17.
Luna, E.; Ton, J. The Epigenetic Machinery Controlling Transgenerational Systemic Acquired Resistance. *Plant Signal. Behav.* **2012,** *7* (6), 615–618.
Lustig, A. J. Hypothesis: Paralog Formation from Progenitor Proteins and Paralog Mutagenesis Spur the Rapid Evolution of Telomere Binding Proteins. *Front. Genet.* **2016,** *7,* 10.
McKay, J. K.; Stinchcombe, J. R. Ecological Genomics of Model Eukaryotes. *Evolution* **2008,** *62* (12), 2953–2957.
Mitchell-Olds, T. Arabidopsis Thaliana and Its Wild Relatives: A Model System for Ecology and Evolution. *Trends Ecol. Evol.* **2001,** *16* (12), 693–700.
Mitchell-Olds, T.; Schmitt, J. Genetic Mechanisms and Evolutionary Significance of Natural Variation in Arabidopsis. *Nature* **2006,** *441* (7096), 947–952.
Mitchell-Olds, T.; Willis, J. H.; Goldstein, D. B. Which Evolutionary Processes Influence Natural Genetic Variation for Phenotypic Traits? *Nat. Rev. Genet.* **2007,** *8* (11), 845–856.
Mittler, R.;Shulaev, V. Functional Genomics, Challenges and Perspectives for the Future. *Physiologia Plantarum* **2013,** *148* (3), 317–321.
Moshelion, M.; Altman, A. Current Challenges and Future Perspectives of Plant and Agricultural Biotechnology. *Trends Biotechnol.* **2015,** *33* (6), 337–342.

Muller, E. E.; Faust, K.; Widder, S.; Herold, M.; Arbas, S. M.;Wilmes, P. Using Metabolic Networks to Resolve Ecological Properties of Microbiomes. *Curr. Opin. Syst. Biol.* **2018,** *8*, 73–80.

Nguyen, D.; Rieu, I.; Mariani, C.; van Dam, N. M. How Plants Handle Multiple Stresses: Hormonal Interactions Underlying Responses to Abiotic Stress and Insect Herbivory. *Plant Mol. Biol.* **2016,** *91* (6), 727–740.

Okushima, Y.; Overvoorde, P. J.; Arima, K.; Alonso, J. M.; Chan, A.; Chan, C.; Theologis, A. Functional Genomic Analysis of the Auxin Response Factor Gene Family Members in Arabidopsis Thaliana: Unique and Overlapping Functions of ARF7 and ARF19. *Plant Cell* **2005,** *17* (2), 444–463.

Ouborg, N. J.;Vriezen, W. H. An Ecologist's Guide to Ecogenomics. *J. Ecol.* **2007,** 8–16.

Pawlowski, J.; Kelly-Quinn, M.; Altermatt, F.; Apotheloloz-Perret-Genetil, L.; Baja, P.; Beggero, A.; Kalhlert, M. The Future of Biotic Indices in the Ecogenomic Era: Integrating (e) DNA Metabarcoding in Biological Assessment of Aquatic Ecosystems. *Sci. Total Environ.* **2018,** *637*, 1295–1310.

Pei, N.; Chen, B.; Kress, W. J. Advances of Community-Level Plant DNA Barcoding in China. *Front. Plant Sci.* **2017,** *8*, 225.

Pinu, F. R.; Beale, D. J.; Paten, A. M.; Kouremenos, k.; Swarup, S.; Schirre, H. J.; Wishart, D. Systems Biology and Multi-Omics Integration: Viewpoints from the Metabolomics Research Community. *Metabolites* **2019,** *9* (4), 76.

Reymond, P.; Weber, H.; Damond, M.; Farmer, E. E. Differential Gene Expression in Response to Mechanical Wounding and Insect Feeding in Arabidopsis. *Plant Cell* **2000,** *12* (5), 707–719.

Rushworth, C. A.; Song, B. H.; Lee, C. R.; Mitchell Olds, T. H. O. M. A. S. Boechera; A Model System for Ecological Genomics. *Mol. Ecol.* **2011,** *20* (23), 4843–4857.

Sardans, J.; Peñuelas, J. Tree Growth Changes with Climate and Forest Type Are Associated with Relative Allocation of Nutrients, Especially Phosphorus, to Leaves and Wood. *Global Ecol. Biogeogr.* **2013,** *22* (4), 494–507.

Schmitz, O. J.; Krivan, V.; Ovadia, O. Trophic Cascades: The Primacy of Trait-Mediated Indirect Interactions. *Ecol. Lett.* **2004,** *7* (2), 153–163.

Schranz, M. E.; Manzaneda, A. J.; Windsor, A. J.; Clauss, M. J.; Mitchell-Olds, T. Ecological Genomics of Boecherastricta: Identification of a QTL Controlling the Allocation of Methionine-vs Branched-Chain Amino Acid-Derived Glucosinolates and Levels of Insect Herbivory. *Heredity* **2009,** *102* (5), 465–474.

Seiler, E.; Trappe, K.; Renard, B. Y. Where Did You Come from, Where Did You Go: Refining Metagenomic Analysis Tools for Horizontal Gene Transfer Characterisation. *PLoS Comput. Biol.* **2019,** *15* (7), e1007208.

Sękiewicz, K.; Sekiewicz, M.; Jasinska, A. K.; Boratynska, K.; Iszkulo, G.; Romo, A. Morphological Diversity and Structure of West Mediterranean Abies Species. *Plant Biosyst. Int. J. Deal. Aspects Plant Biol.* **2013,** *147* (1), 125–134.

Simmler, C.; Napolitano, J. G.; McAlpine, J. B.; Chen, S. N.; Pauli, G. F. Universal Quantitative NMR Analysis of Complex Natural Samples. *Curr. Opin. Biotechnol.* **2014,** *25*, 51–59.

Snoeren, T. A.; De Jong, P. W.;Dicke, M. Ecogenomic Approach to the Role of Herbivore-Induced Plant Volatiles in Community Ecology. *J. Ecol.* **2007,** *95* (1), 17–26.

Tagu, D.; Colbourne, J. K.; Nègre, N. Genomic Data Integration for Ecological and Evolutionary Traits in Non-Model Organisms. *BMC Genom.* **2014,** *15* (1), 1–16.

Taylor, G.; Street, N. R.; Tricker, P. J.; Sjödin, A.; Graham, L.; Skogström, O.; Jansson, S. The Transcriptome of Populus in Elevated CO_2. *New Phytol.* **2005,** *167* (1), 143–154.

Thiellement, H.; Bahrman, N.; Damerval, C.; Plomion, C.; Rossignol, M.; Santoni, V.; Zivy, M. Proteomics for Genetic and Physiological Studies in Plants. *Electrophor. Int. J.* **1999,** *20* (10), 2013–2026.

Thiellement, H.; Zivy, M.;Plomion, C. Combining Proteomic and Genetic Studies in Plants. *J. Chromatogr. 2002B, 782* (1–2), 137–149.

Tian, D.; Traw, M. B.; Chen, J. Q.; Kreitman, M.;Bergelson, J. Fitness Costs of R-Gene-Mediated Resistance in Arabidopsis Thaliana. *Nature* **2003,** *423* (6935), 74–77.

Timmermans, M. J. T. N.; Roelofs, D.; Mariën, J.; Van Straalen, N. M. Revealing Pancrustacean Relationships: Phylogenetic Analysis of Ribosomal Protein Genes Places Collembola (Springtails) in a Monophyletic Hexapoda and Reinforces the Discrepancy between Mitochondrial and Nuclear DNA Markers. *BMC Evol. Biol.* **2008,** *8* (1), 1–10.

Tiwari, P.; Bajpai, M.; Singh, L. K.; Mishra, S.; Yadav, A. N. Phytohormones Producing Fungal Communities: Metabolic Engineering for Abiotic Stress Tolerance in Crops. In *Agriculturally Important Fungi for Sustainable Agriculture*; Springer: Cham, 2020; pp 171–197.

Tripathi, A. M.; Tyagi, A.; Kumar, A.; Singh, A.; Singh, S.; Chaudhary, L. B.; Roy, S. The Internal Transcribed Spacer (ITS) Region and trnhH-psbA Are Suitable Candidate Loci for DNA Barcoding of Tropical Tree Species of India. *PloS One* **2013,** *8* (2), e57934.

Ungerer, M. C.; Loretta, C. J.; Michael, A. H. Ecological Genomics: Understanding Gene and Genome Function in the Natural Environment. *Heredity* **2008,** *100* (2), 178–183.

Van Aggelen, G.; Ankley, G. T.; Baldwin, W. S.; Bearden, D. W.; Benson, W. H.; Chipman, J. K.; Falciani, F. Integrating Omic Technologies into Aquatic Ecological Risk Assessment and Environmental Monitoring: Hurdles, Achievements, and Future Outlook. *Environ. Health Perspect.* **2010,** *118* (1), 1–5.

Van Dam, N. M. How Plants Cope with Biotic Interactions. *Plant Biol.* **2009,** *11* (1), 1–5.

Van Dam, N. M.; Poppy, G. M. Why Plant Volatile Analysis Needs Bioinformatics–Detecting Signal from Noise in Increasingly Complex Profiles. *Plant Biol.* **2008,** *10* (1), 29–37.

Vescio, R.; Abenavoli, M. R.; Sorgonà, A. Single and Combined Abiotic Stress in Maize Root Morphology. *Plants* **2021,** *10* (1), 5.

Weider, L. J.; Elser, J. J.; Crease, T. J.; Mateos, M.; Cotner, J. B.;Markow, T. A. The Functional Significance of Ribosomal (r) DNA Variation: Impacts on the Evolutionary Ecology of Organisms. *Annu. Rev. Ecol. Evol. Syst.* **2005,** *36*, 219–242.

Wolfender, J. L.; Marti, G.; Thomas, A.; Bertrand, S. Current Approaches and Challenges for the Metabolite Profiling of Complex Natural Extracts. *J. Chromatogr. A* **2015,** *1382*, 136–164.

Xia, X. DAMBE7: New and Improved Tools for Data Analysis in Molecular Biology and Evolution. *Mol. Biol. Evol.* **2018,** *35* (6), 1550–1552.

Index

W

Single-Molecule Tools for Bioanalysis

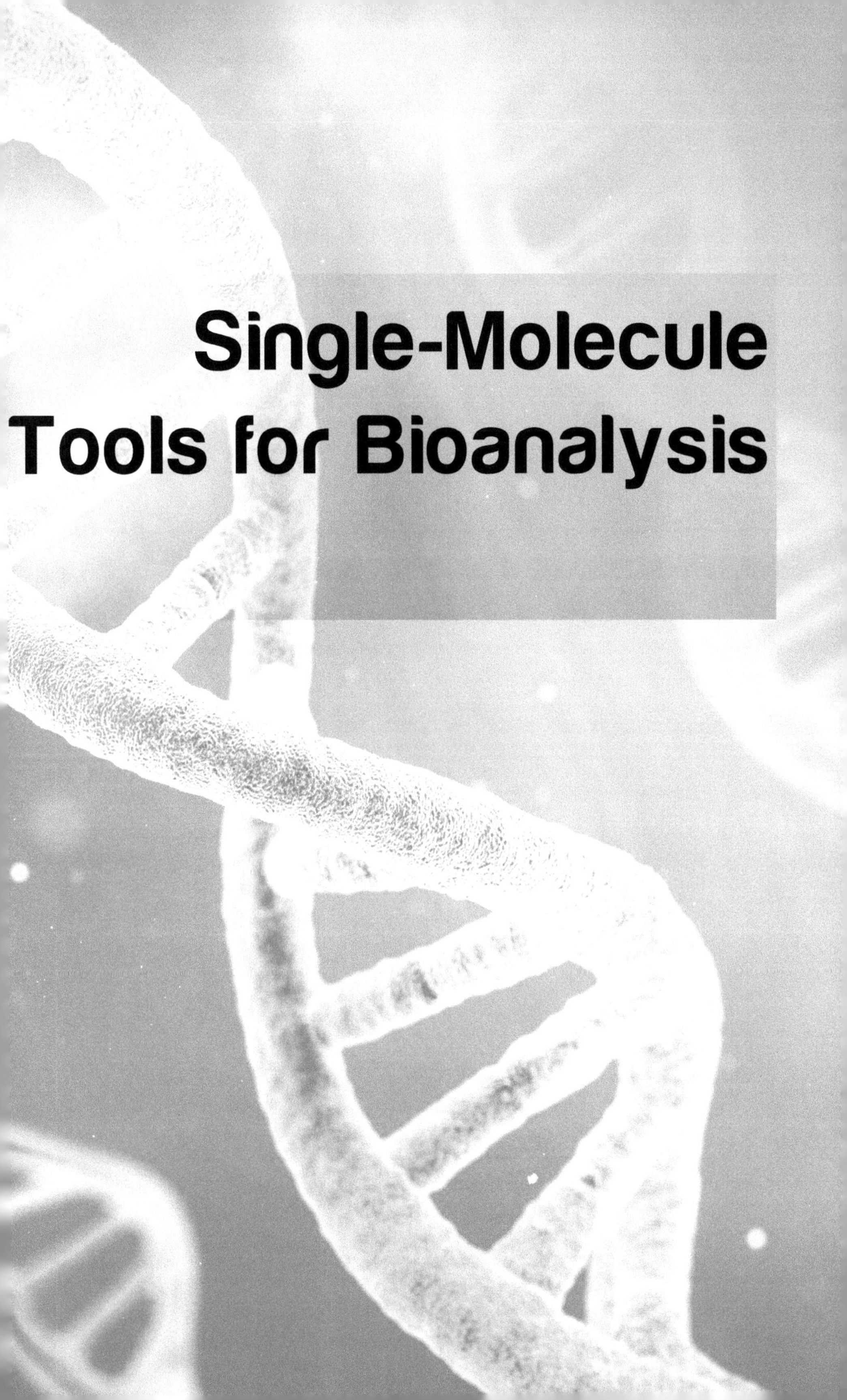

Single-Molecule Tools for Bioanalysis

edited by
Shuo Huang

JENNY STANFORD
PUBLISHING

Published by

Jenny Stanford Publishing Pte. Ltd.
Level 34, Centennial Tower
3 Temasek Avenue
Singapore 039190

Email: editorial@jennystanford.com
Web: www.jennystanford.com

British Library Cataloguing-in-Publication Data
A catalogue record for this book is available from the British Library.

Single-Molecule Tools for Bioanalysis

ISBN 978-981-4800-44-0 (Hardcover)
ISBN 978-1-003-18913-8 (eBook)